GLUTATHIONE IN HEALTH AND DISEASE

Edited by **Pinar Erkekoglu** and **Belma Kocer-Gumusel**

Glutathione in Health and Disease

http://dx.doi.org/10.5772/intechopen.70965

Edited by Pinar Erkekoglu and Belma Kocer-Gumusel

Contributors

Juliana Echevarria-Lima, Hongli Wu, Christy Xavier, Xiaobin Liu, Duen-Shian Wang, Yang Liu, Angela Adamski Da Silva Reis, Laura Raniere Borges dos Anjos, Ana Cristina Silva Rebelo, Gustavo Rodrigues Pedrino, Rodrigo Da Silva Santos, Roberto Burini, Hugo Kano, Yong-Ming Yu, Pinar Erkekoglu

Notice

Statements and opinions expressed in the chapters are these of the individual contributors and not necessarily those of the editors or publisher. No responsibility is accepted for the accuracy of information contained in the published chapters. The publisher assumes no responsibility for any damage or injury to persons or property arising out of the use of any materials, instructions, methods or ideas contained in the book.

First published in London, United Kingdom, 2018 by IntechOpen

IntechOpen is the global imprint of INTECHOPEN LIMITED, registered in England and Wales, registration number: 11086078, The Shard, 25th floor, 32 London Bridge Street

London, SE19SG – United Kingdom

Printed in Croatia

British Library Cataloguing-in-Publication Data

A catalogue record for this book is available from the British Library

Additional hard copies can be obtained from orders@intechopen.com

Glutathione in Health and Disease, Edited by Pinar Erkekoglu and Belma Kocer-Gumusel

p. cm.

Print ISBN 978-1-78984-275-3

Online ISBN 978-1-78984-276-0

We are IntechOpen,
the world's leading publisher of
Open Access books
Built by scientists, for scientists

3,800+
Open access books available

116,000+
International authors and editors

120M+
Downloads

Our authors are among the

151
Countries delivered to

Top 1%
most cited scientists

12.2%
Contributors from top 500 universities

CLARIVATE ANALYTICS
BOOK
CITATION
INDEX
INDEXED

WEB OF SCIENCE™

Selection of our books indexed in the Book Citation Index
in Web of Science™ Core Collection (BKCI)

Interested in publishing with us?
Contact book.department@intechopen.com

Numbers displayed above are based on latest data collected.
For more information visit www.intechopen.com

Meet the editors

Pınar Erkekoglu was born in Ankara, Turkey. She graduated from Hacettepe University Faculty of Pharmacy (BS). Later, she received her MSci and PhD degrees in toxicology. She completed a part her PhD studies in Grenoble, France in Universite Joseph Fourier and CEA/INAC/LAN after receiving full scholarship from both Erasmus Scholarship Program and from CEA. She worked as a post-doc and a visiting associate in MIT Biological Engineering Department. She is working as an associate professor in Hacettepe University Faculty of Pharmacy Department of Pharmaceutical Toxicology since 2013. Her main interests are endocrine disrupting chemicals, oxidative stress and aromatic amines. She has published more than 140 papers in national and international journals. She is a European Registered Toxicologist (ERT) since 2014.

Belma Kocer-Gumusel graduated from Hacettepe University Faculty of Pharmacy. She received her PhD degree in toxicology in 1999. She worked as a postdoctoral fellow in Joseph Fourier University and CEA. She served as a faculty staff and retired as a full professor from Hacettepe University Faculty of Pharmacy. She now serves as the department head in Lokman Hekim University Faculty of Pharmacy Department of Pharmaceutical Toxicology. Her main interests are endocrine diseases, oxidative stress and selenium. She has published more than 150 papers in national and international journals. She is a member of national and international scientific societies, including Turkish Society of Toxicology, EUROTOX and IUTOX. She is a European Registered Toxicologist (ERT) since 2011.

Contents

Preface

Reduced glutathione (GSH) is a water-soluble tripeptide with the structure of g-glutamyl-cysteinyl-glycine. GSH is the most important thiol in living organisms. It is the key component of antioxidant system, serves as a free radical scavenger and can be effective against the attacks of many reactive species, including electrophilic substances and epoxides. Upon contact with reactive species, GSH is converted to oxidized glutathione (GSSG) by glutathione peroxidases and later can be reduced to GSH again by glutathione reductase. Therefore, there is a cycle of GSH and this cycle provides higher intracellular levels of GSH. GSH levels are depleted to different degrees if the cellular death mechanisms are triggered. GSH depletion and apparent oxidative stress may also cause cytotoxicity. GSH was shown to be preventive against aging, cancer, heart disease and dementia. Moreover, GSH supplementation (mainly as cysteine) can help to reduce the symptoms of many diseases and can be beneficial in different conditions like autism, Alzheimer's disease, Parkinson's disease, HIV/AIDS, hepatitis, type 2 diabetes, cystic fibrosis and certain infections. This book is mainly focused on GSH in health and disease. The readers will receive information on the diverse functions of GSH and GSH-related enzymes, the importance of GSH status against oxidative stress and the interaction between GSH and nervous system-related infections. We believe that readers will get qualified scientific knowledge and a general overview of the importance of GSH and GSH-related enzymes in health and in different pathological conditions from this book.

Assoc.Prof.Dr. Pinar Erkekoglu
Hacettepe University
Faculty of Pharmacy
Department of Toxicology
Ankara, Turkey

Prof.Dr. Belma Kocer-Gumusel
Hacettepe University
Faculty of Pharmacy
Department of Toxicology
Ankara, Turkey

Glutathione - Evolution and Functions

Introductory Chapter: A General Overview of Glutathione, Glutathione Transport, and Glutathione Applications

Pinar Erkekoglu

Additional information is available at the end of the chapter

http://dx.doi.org/10.5772/intechopen.81594

1. Introduction

Reduced glutathione (GSH) is a water-soluble tripeptide with the structure of γ-glutamyl-cysteinyl-glycine. The gamma bond between glutamic acid and cysteine provides stability to GSH as there are lower amounts of γ-peptidases in biological systems when compared to α-peptidases [1].

GSH is the most important thiol in living organisms. GSH is a catalyst, reductant, and reactant. This molecule is found in large quantities (millimolar concentrations) in organs exposed to toxins such as liver, kidney, lungs, and intestines. However, in body fluids, GSH concentrations are at micromolar concentrations [2].

GSH is synthesized in the cell cytosol. In a reaction catalyzed by "γ-glutamylcysteine synthetase," L-cysteine and L-glutamate form "γ-glutamylcysteine." By the addition of glycine, GSH is formed in a reaction catalyzed by "glutathione synthetase." The catabolism of GSH is mainly catalyzed by "γ-glutamyl transpeptidase" (forming glutamate and cysteinylglycine) and "dipeptidases" (forming cysteine and glycine). Cysteine is then catabolized to mercapturic acid [3].

GSH plays a role in amino acid transport, protein synthesis, DNA synthesis and protection, and more generally, in cellular detoxification. The main source of plasma GSH is liver. As part of their physiological functions, viable cells such as hepatocytes and macrophages extrude this particular thiol. This phenomenon supplies antioxidant protection for the extracellular environment as well [4].

GSH acts essentially as an intracellular key component of antioxidant system, serves as a free radical scavenger and can be effective against the attacks of many reactive species, including electrophilic substances and epoxides. Upon contact with reactive species, GSH is converted to oxidized glutathione (GSSG, dimeric glutathione) by glutathione peroxidases (GPxs) and later can be reduced to GSH again by glutathione reductase (GR). Therefore, there is a cycle of GSH and this cycle provides higher intracellular levels of GSH. On the other hand, glutathione S-transferases (GSTs) can form conjugates between GSH and endogenous (e.g., estrogens) or exogenous substances (e.g., electrophiles like arene oxides, organic halides, or unsaturated carbonyls). The decrease in the activities of GSTs may increase risk for disease; however, some GSH conjugates can also be toxic, paradoxically [5].

The transport of GSH through plasma membrane is regulated by a switch mechanism orchestrated by open/closed configuration of the transporters. This transfer from the cell to the extracellular environment occurs according to a concentration gradient. The transport is uniport and cells usually export GSH rather than import as intracellular GSH levels are higher than extracellular fractions [6].

The molecular nature of GSH transporters is still elusive though these transports are functionally identified as sinusoidal or canalicular type due to their position in the hepatic anatomy and their responsiveness to specific inhibitors [7, 8]. Data from different reports show that GSH transporters coincide with the multi-drug resistance-associated proteins (MRPs) [6, 9]. The regulatory mechanism/s behind the activity of GSH transporters is/are still ambiguous. These transports are possibly being controlled by the differential concentration of GSH on the internal vs. external side of the cellular membranes. Thereby, their control may mainly rely on the zonal control of GSH levels by intracellular trafficking [4]. A different regulation mechanism operates in the export of GSSG when cells are subject to oxidative stress or when GSSG cannot be reduced to GSH by GR. Even though GSSG was shown to be target of MRP [6], in conditions of oxidative stress GSSG crosses the plasma membrane and passively exits from cells. This is a balancing mechanism which helps to avoid a dangerous drop in the redox (GSH/GSSG) ratio due to the accumulation of GSSG, and the consequent redox imbalance [6, 10].

It is still a question mark whether GSH depletion is a cause or an outcome of different pathological conditions and exposure to certain environmental chemicals. Several diseases may cause depletion of intracellular GSH levels. On the other hand, environmental chemicals and drugs may also lead to GSH repression. In addition, GSH levels are depleted at different degrees if the cellular death mechanisms are triggered by different chemicals or conditions. GSH depletion and apparent oxidative stress may also cause cytotoxicity [10–12].

Reduced glutathione was shown to be preventive against aging, cancer, heart disease, and dementia. Moreover, GSH supplementation (mainly as cysteine) can help to reduce the symptoms of many diseases and can be beneficial in different conditions like autism, Alzheimer's disease, Parkinson's disease, human immunodeficiency virus (HIV)/acquired immunodeficiency syndrome (AIDS), hepatitis, type 2 diabetes, cystic fibrosis, and certain infections [13–16].

Plasma and liver GSH levels were shown to decrease dramatically in certain liver disease patients (i.e., viral hepatitis, chronic hepatitis, chronic liver injury, and liver cirrhosis) [17]. GSH also plays an important role in the activation of T-lymphocytes [18]. In HIV infection, a systemic drop in intracellular/extracellular GSH is linked to an increase of virus replication. Moreover, cysteine deficiency may also lead to decrease in GSH levels which also leads to high viral load [18, 19].

In Parkinson's disease, the level of glutathione in *substantia nigra* was found to be lower in Parkinsonian patients compared to controls. This decrease may be related to the increased degradation of GSH by γ-glutamyltranspeptidase [20–23]. GSH levels were also found to be depleted in certain lung diseases, like acute respiratory disease (ARDS), and in neonatal lung damage [24–26].

A reduction of GSH levels was determined in the ischemic tissue during myocardial ischemia and reperfusion and myocardial injury was found to be negatively associated with the myocardial concentration of GSH. The administration of γ-glutamylcysteine or N-acetyl cysteine (NAC) markedly reduced the infarct size and myocyte death [27–30]. GSH levels were reported to be significantly reduced in renal ischemia and in cyclosporin intoxication as both conditions induce lipid peroxidation in microsomes [31–34].

Studies on the interaction between GSH levels and aging are still contradictory. Therefore, large-scale epidemiological studies are needed in order to reach conclusions [35–37]. On the other hand, GSH was shown to protect against several types of cancer [35–37]. For instance, administration of GSH can provide decrease in the rate of different types of cancers [38, 39]. However, GSH treatment should be reconsidered in some types of cancer as GSH is a double-edged knife in cancer treatment and may lead to the development of resistance to chemotherapy [38, 39].

GSH has poor bioavailability. This phenomenon restricts its direct use in clinics [30]. Hydrophobic forms such as monoethylester of GSH have been synthesized to overcome this restriction. These synthetic forms are cleaved by cellular esterases to form GSH. After such forms are administered to after GSH-deprived rats by oral route, an increase in GSH concentrations was observed in both plasma and liver [40]. Due to the oxidation of cystine, cysteine may cause toxicity. Therefore, N-acetyl cysteine (NAC) has been used as an exogenous source of cysteine to provide intracellular glutathione in GSH-deficient patients. After hydrolysis, NAC can be a source of cysteine, the major amino acid in synthesis of GSH [41]. The administration of NAC to HIV positive patients provides increases in GSH levels of CD4+ lymphocytes, inhibits the activity of nuclear factor kappa B (NFkB), and arrests viral replication [42].

In conclusion, GSH continues to be investigated in diverse pathological conditions. Cancer, liver diseases, neuropathological diseases, acute respiratory distress syndrome, HIV/AIDS, and aging are the main research fields for studies on GSH. However, there is a long road ahead in order to use GSH or its different forms in clinics for certain conditions. GSH will continue to be the subject of new studies and hopefully GSH or its different forms will be used as a drug or adjunct therapy for certain pathological conditions in the future.

Author details

Pinar Erkekoglu

Address all correspondence to: erkekp@yahoo.com

Department of Toxicology, Hacettepe University Faculty of Pharmacy, Ankara, Turkey

References

[1] Page MJ, Di Cera E. Evolution of peptidase diversity. The Journal of Biological Chemistry. 2008;**283**(44):30010-30014

[2] Kidd PM. Glutathione: Systemic protectant against oxidative and free radical damage. Alternative Medicine Review. 1997;**1**:155-176

[3] Gaté L, Paul J, Ba GN, Tew KD, Tapiero H. Oxidative stress induced in pathologies: The role of antioxidants. Biomedicine & Pharmacotherapy. 1999;**53**(4):169-180

[4] De Nicola M, Ghibelli L. Glutathione depletion in survival and apoptotic pathways. Frontiers in Pharmacology. 2014;**5**:267

[5] Birben E, Sahiner UM, Sackesen C, Erzurum S, Kalayci O. Oxidative stress and antioxidant defense. World Allergy Organization Journal. 2012;**5**(1):9-19

[6] Ballatori N, Krance SM, Marchan R, Hammond CL. Plasma membrane glutathione transporters and their roles in cell physiology and pathophysiology. Molecular Aspects of Medicine. 2009;**30**(1-2):13-28

[7] Yi JR, Lu S, Fernández-Checa J, Kaplowitz N. Expression cloning of the cDNA for a polypeptide associated with rat hepatic sinusoidal reduced glutathione transport: Characteristics and comparison with the canalicular transporter. Proceedings of the National Academy of Sciences of the United States of America. 1995;**92**(5):1495-1499

[8] Bachhawat AK, Thakur A, Kaur J, Zulkifli M. Glutathione transporters. Biochimica et Biophysica Acta. 2013;**1830**(5):3154-3164

[9] Franco R, Cidlowski JA. Glutathione efflux and cell death. Antioxidants & Redox Signaling. 2012;**17**(12):1694-1713

[10] Jozefczak M, Remans T, Vangronsveld J, Cuypers A. Glutathione is a key player in metal-induced oxidative stress defenses. International Journal of Molecular Sciences. 2012;**13**(3):3145-3175

[11] Mytilineou C, Kramer BC, Yabut JA. Glutathione depletion and oxidative stress. Parkinsonism & Related Disorders. 2002;**8**(6):385-387

[12] Ahmad S. Oxidative stress from environmental pollutants. Archives of Insect Biochemistry and Physiology. 1995;**29**(2):135-157

[13] Uttara B, Singh AV, Zamboni P, Mahajan RT. Oxidative stress and neurodegenerative diseases: A review of upstream and downstream antioxidant therapeutic options. Current Neuropharmacology. 2009;**7**(1):65-74

[14] Lapenna D, Ciofani G, Calafiore AM, Cipollone F, Porreca E. Impaired glutathionerelated antioxidant defenses in the arterial tissue of diabetic patients. Free Radical Biology & Medicine. 2018;**124**:525-531

[15] Morris D, Khurasany M, Nguyen T, Kim J, Guilford F, Mehta R, et al. Glutathione and infection. Biochimica et Biophysica Acta. 2013;**1830**(5):3329-3349

[16] Hudson VM. New insights into the pathogenesis of cystic fibrosis: Pivotal role of glutathione system dysfunction and implications for therapy. Treatments in Respiratory Medicine. 2004;**3**(6):353-363

[17] Yuan L, Kaplowitz N. Glutathione in liver diseases and hepatotoxicity. Molecular Aspects of Medicine. 2009;**30**(1-2):29-41

[18] FJT S, Roederer M, Israelski M, Dubp J, Mole LA, McShane D, et al. Intracellular glutathione levels in T-cell subsets decrease in HIV-infected individuals. AIDS Research and Human Retroviruses. 1992;**8**:305-311

[19] Holroyd KJ, Buhl R, Borok Z, Roum JH, Bokser AD, Grimes GJ, et al. Correction of glutathione deficiency in the lower respiratory tract of HIV seropositive individuals by glutathione aerosol treatment. Thorax. 1993;**48**(10):985-989

[20] Jenner P. Oxidative damage in neurodegenerative disease. Lancet. 1994;**344**(8925):796-798

[21] Smeyne M, Smeyne RJ. Glutathione metabolism and Parkinson's disease. Free Radical Biology & Medicine. 2013;**62**:13-25

[22] Dias V, Junn E, Mouradian MM. The role of oxidative stress in Parkinson's disease. Journal of Parkinson's Disease. 2013;**3**(4):461-491

[23] Sian J, Dexter DT, Lees AJ, Daniel S, Jenner P, Marsden CD. Glutathione-related enzymes in brain in Parkinson's disease. Annals of Neurology. 1994;**36**(3):356-361

[24] Christofidou-Solomidou M, Muzykantov VR. Antioxidant strategies in respiratory medicine. Treatments in Respiratory Medicine. 2006;**5**(1):47-78

[25] Pachter P, Timerman AP, Lykens MG, Merola AJ. Deficiency of alveolar fluid glutathione in patients with sepsis and the adult respiratory distress syndrome. Chest. 1991;**100**: 1397-1403

[26] Grigg J, Barber A, Silverman M. Bronchoalveolar lavage fluid glutathione in incubated premature infants. Archives of Disease in Childhood. 1993;**69**:49-51

[27] Glantzounis GK, Yang W, Koti RS, Mikhailidis DP, Seifalian AM, Davidson BR. The role of thiols in liver ischemia-reperfusion injury. Current Pharmaceutical Design. 2006; **12**(23):2891-2901

[28] Ferrari R, Guardigli G, Mele D, Percoco GF, Ceconi C, Curello S. Oxidative stress during myocardial ischaemia and heart failure. Current Pharmaceutical Design. 2004;**10**(14): 1699-1711

[29] Forman MB, Puett DW, Cates CU, MC Croskey DE, Beckman JK, Greene HL, et al. Glutathione redox pathway and perfusion injury. Circulation. 1988;**78**:202-203

[30] Exner R, Wessner B, Manhart N, Roth E. Therapeutic potential of glutathione. Wiener Klinische Wochenschrift. 2000;**112**(14):610-616

[31] Wang C, Salahudeen AK. Cyclosporine nephrotoxicity: Attenuation by an antioxidant-inhibitor of lipid peroxidation in vitro and in vivo. Transplantation. 1994;**58**:940-946

[32] Duruibe V, bkonmah A, Blyden GT. Effect of cyclosporin on rat liver and kidney glutathione content. Pharmacology. 1989;**39**:205-212

[33] Yang HY, Lee TH. Antioxidant enzymes as redox-based biomarkers: A brief review. BMB Reports. 2015;**48**(4):200-208

[34] Weinberg JM. The cell biology of ischemic renal injury. Kidney International. 1991; **39**(3):476-500

[35] Homma T, Fujii J. Application of glutathione as anti-oxidative and anti-aging drugs. Current Drug Metabolism. 2015;**16**(7):560-571

[36] Sekhar RV, Patel SG, Guthikonda AP, Reid M, Balasubramanyam A, Taffet GE, et al. Deficient synthesis of glutathione underlies oxidative stress in aging and can be corrected by dietary cysteine and glycine supplementation. The American Journal of Clinical Nutrition. 2011;**94**(3):847-853

[37] Go YM, Jones DP. Redox theory of aging: Implications for health and disease. Clinical Science (London, England). 2017;**131**(14):1669-1688

[38] Traverso N, Ricciarelli R, Nitti M, Marengo B, Furfaro AL, Pronzato MA, et al. Role of glutathione in cancer progression and chemoresistance. Oxidative Medicine and Cellular Longevity. 2013;**2013**:972913

[39] Bansal A, Simon MC. Glutathione metabolism in cancer progression and treatment resistance. The Journal of Cell Biology. 2018;**217**(7):2291-2298

[40] Grattagliano I, Wieland P, Schranz C, Lauterburg BH. Disposition of glutathione monoethyl ester in the rat: Glutathione ester is a slow release of extracellular glutathione. The Journal of Pharmacology and Experimental Therapeutics. 1995;**272**:484-488

[41] Lomaestro BM, Malone M. Glutathione in health and disease: Pharmacotherapeutic issues. The Annals of Pharmacotherapy. 1995;**29**(12):1263-1273

[42] Mihm S, Ennen J, Pessara U, Kurth R, Droge W. Inhibition of HIV-1 replication and NFkappaB activity by cysteine and cysteine derivatives. AIDS. 1991;**5**:497-503

The Life Evolution on the Sulfur Cycle: From Ancient Elemental Sulfur Reduction and Sulfide Oxidation to the Contemporary Thiol-Redox Challenges

Roberto C. Burini, Hugo T. Kano and Yong-Ming Yu

Additional information is available at the end of the chapter

http://dx.doi.org/10.5772/intechopen.76749

Abstract

Organismal evolution led to innovations in metabolic pathways, many of which certainly modified the surface chemistry of the Earth. Volcanic activity introduced inorganic compounds (H_2, CO_2, CH_4, SO_2, and H_2S) driving the metabolism of early organisms of the domains archaea and bacteria. In the absence of light, H_2S and Fe^{2+} would have been the major electron donors and the electron acceptors could be either oxidized species such as the sulfurs, sulfate, and elemental sulfur, or carbon dioxide by the fermentation of acetate (forming methane). Elemental sulfur was produced by the reaction between H_2S and SO_2, while anoxygenic photosynthesis may have provided the sulfate which removed oceanic ferrous iron by its precipitation as sulfide into sediments. Hence, the sulfur cycle participation in life evolution comes from ancient anoxygenic elemental sulfur reduction generating environmental sulfide incorporated as mitochondrial Fe-S for the electron-transport chains. Anoxygenic photosynthesis may have provided the necessary sulfate to promote the evolution of sulfate-reducing bacteria. The evolution of oxygenic photosynthesis provided for diverse metabolic possibilities including non-photosynthetic sulfide oxidation, nitrification, and methanotrophy. An increase in oxygen levels would account for oxidative sulfur cycle, evolution of colorless sulfur bacteria, and emergence of large multicellular animals. Oxygen, initially a waste product of photosynthesis, first reacted with sulfur, iron or methane and latter accumulated in atmosphere resulting in more carbon production. Oxygenic photosynthesis becomes a positive feedback on the oxidation of the Earth-surface environment causing the growth and stabilization of continental platforms and carbon burial with more atmosphere oxidation. An increase in oxygen levels would account for oxidative sulfur cycle, evolution of colorless sulfur bacteria, and emergence of large multicellular animals. Oxygen enabled more efficient energy transformation from dietary food to ATP. However, evolution for mammals living on dry land has been closely linked to the adaptation of changes in O_2 concentration in the environment, which means mitochondrial aerobic respiration. By using ancestral geochemistry of iron-sulfur clusters at the protein complexes I and II, the respiratory chains become badly insulated wires in the presence of oxygen (with reduced respiratory complexes) and there is leakage of electrons on to molecular oxygen. The electron leakage results in the formation of superoxide anion (SO) that remains within the mitochondrial matrix. If not promptly detoxified by anti-oxidative defenses, SO and its derived-oxidative species can alter cell signaling or attack cell structures leading to cell apoptosis. Sulfur-containing compounds participate either in oxidative stress generation (at endoplasmic reticulum) or in (thiol) antioxidant defenses (mainly glutathione), thus functioning as redox sensing for enzyme activity and gene expression. Sulfur compounds that contributed for electron leakage and oxidative stress have counteractions by thiol participation either as antioxidant defensors and/or as redox-modulators or cell functions, influencing life evolution and contemporary diseases.

Keywords: life evolution, evolutive sulfur cycle, evolutive oxygen, oxidative metabolism, thiol-redox

---

1. Introduction

1.1. The life evolution on the sulfur cycle

Organismal evolution led to innovations in metabolic pathways, many of which certainly modified the surface chemistry of the Earth. Such changes in surface chemistry provided new metabolic opportunities that promoted further evolutionary innovations (**Figure 1**). Life on Earth evolved under anaerobic conditions, with metabolic pathways centered on nitrogen, sulfur and carbon. There are two principal avenues of inquiry relevant to reconstructing the history of the sulfur cycle. One avenue relies on the comparison of molecular sequences derived from biologically essential proteins and genetic materials. Other is the geologic record that can provide direct evidence for the state of chemical oxidation of the Earth-surface, with possible indications of when specific bacterial metabolisms first occurred. However, the most complete understanding of the course of the chemical evolution of the surface environment will fully integrate organismal phylogenies with geochemical and geological evidence for surface change [1].

1.2. Evolutionary phylogenies

A deeper understanding of the evolutionary relationship among organisms is possible from phylogenies derived from comparisons of the small subunit (SSU) of the ribosomal RNA molecule (rRNA) comprising the 16S subunit for prokaryotic organisms and 18S subunit for eukaryotes [2]. Sequence analysis of small subunit of rRNA has revealed that all life can be divided into three principal domains; these are the bacteria, the archaea, and the eucarya [3].

The small subunit of rRNA-based tree of life bears little resemblance to the "classic" tree of life which divided the living world into five kingdoms [4]: Animalia, Plantae, Fungi, Protista, and Monera (prokaryotes). However, the new tree of life emphasizes the genetic diversity of prokaryotes (bacteria and archaea) and shows that the history of life on Earth is largely a history of prokaryotic evolution, not the evolution of macroscopic organisms as previous phylogenetic

3.8-4.0 Ga	Large continental land not yet formed Sub-aqueous and sub-aerial vigorous vulcanic activity generating inorganic compounds: H_2, CO_2, CH_3, SO_2, H_2S Virtually oxygen and far higher levels of CO_2 (acidic oceans) Ecosystem: thermophilic - hyperthermophilic Ergogenesis: anoxygenic phototrophs, fermentative, heterotrophs Metabolism: chemoheteroytophic (absence of light) and chemolithoantotrophic using O_2, NO_3 and metal oxide as electron acceptor Organisms: forms of life metabolizing sulfur compounds: procaryotic of Archea domains and bacteria
3.4 - 3.5 Ga	Anoxygenic photoseynthesis established Ecosystem: sulfur compounds accumulation, e.g. sulfur waste from H_2S elemental reduction and, sulfate waste from sulfite reduction Organism metabolism: a)anoxygenic phototrophic bacteria (oxidize) H_2S to elemental sulfur and sulfate, b)chemolithoautotrophic anoxygenic phototrophic (green sulfur bacteria), c)chemolithoautotrophic oxygenic phototrophic (cyanobacteria, purple bacteria, Gram + bacteria)
2.8 Ga	First oxygenic photosynthesis Ecosystem: increased carbon production and oxidation Metabolism: sunlight electrolyze of water and electron transferred to CO2 to form sugars
2.3 Ga	Ocean sulfate reached 1 mM
1.8 Ga	Sulfate reduction rates exceeded the deliver of iron and thus, sulfite-ferrous precipitation stopped in oceans
0.75 Ga	Ocean sulfate accumulation beyond 1 mM
0.54 Ga	Atmospheric oxygen achieved the present levels Second major burial episodes of organic matter

Figure 1. Timeline of life sulfur and oxygen cycle.

schemes would suggest [1]. Near the root of the tree of life are numerous bacteria metaboliz-ing sulfur species including organisms living from dissimilatory elemental sulfur reduction, dissimilatory sulfate reduction, and anoxygenic photosynthesis. These metabolisms are likely very ancient (**Figure 1**).

Anoxygenic photosynthesis may have provided the necessary sulfate to promote the evolution of sulfate-reducing bacteria. Furthermore, the evolution of oxygenic photosynthesis provided for diverse metabolic possibilities including non-photosynthetic sulfide oxidation, nitrification and methanotrophy, Hence, it is likely that, in one way or another, some of the earliest organ-isms on Earth gained energy from the metabolism of sulfur compounds (**Figure 1**).

1.3. Evolutionary sulfur ecosystems

The geological record begins in the early Archean (3.8–3.9 Ga) (gyga annum = billion years). Large continental land masses had probably not yet formed and the rock record records vigor-ous volcanic activity in both sub-aerial and sub-aqueous settings [5–7]. A primitive early Earth terrestrial ecosystem was likely thermophilic to hyperthermophilic and housed around active hydrothermal areas with anoxygenic photosynthesis (when light was available) producing organic matter and oxidized sulfur species (**Figure 1**). In the absence of light (chemohetero-trophic metabolisms), the organic matter production utilized H_2 as the electron donor and oxidized species as the electron acceptors. Hence, the oxidized sulfur species could have been used as electron acceptors in the mineralization of organic matter, completing the carbon cycle. In chemoheterotrophic metabolism, organic compounds were oxidized by the reduction of elemental sulfur and sulfate (forming H_2S) as well as by the carbon dioxide from fermentation of acetate (forming methane) [1].

In the anoxygenic photosynthesis, photo-oxidation of chlorophyll transforms it into an oxidant, which can strip electrons from many sources, passing them ultimately onto CO_2. If H_2S is the electron donor, the waste product is sulfur [1]. Anoxygenic photosynthesis by volcanogenic constituents includes sulfate reduction and elemental sulfur reduction. Elemental sulfur was produced by the reaction between H_2S and SO_2. A possible sulfate source might have been the hydrolysis of volcanogenic SO_2 originating from relatively oxidized magmas [8]. But seems, anoxygenic photosynthesis is the most important source of sulfate (**Figure 1**).

Early Earth had an anoxic atmosphere, with an anoxic ocean containing low concentrations of sulfate. By 3.5 Ga, anoxygenic photosynthesis was established and provided a weak source of sulfate to the global ocean. The origin of that sulfate was attributed to the anoxygenic phototrophic oxidation of primary mantle-derived sulfide to sulfate. However, in 3.4 Ga, there was the accumulation of minimally fractionated evaporitic sulfates in association with magmatically derived sulfides. The stable isotope record of sedimentary sulfides indicates that sulfate first accumulated into the global ocean to concentrations approx. 1 mM at around 2.3 Ga (**Figure 1**). At low concentrations, the rates of delivery of sulfate into sediments may be severely limited by diffusion of sulfate across the sediment-water interface.

Throughout the Archean and early Proterozoic the deep oceans contained appreciable concentrations of dissolved ferrous iron, and banded iron formations were a common form of chemical sediment. When the production rate of sulfide exceeded the delivery flux of iron, dissolved iron was removed from the oceans by reaction with sulfide. As a consequence, the oceans became sulfidic. At this point, sulfide accumulated and precipitated ferrous iron from the solution. It is suggested that banded iron formations (with sulfide) were stopped forming (about 1.8 Ga) when sulfate levels increased and, consequently, sulfate reduction rates rose to the point of exceeding the delivery flux of iron to the oceans. A rise in sulfate reduction rates would be promoted by an increase in the sulfate concentration beyond 1 mM providing a higher flux of sulfate into sediments [1]. It is proposed that the oceans remained sulfide-rich until the Neoproterozoic, where renewed deposition of banded iron formations occurred at around 0.75 Ga (**Figure 1**). The competing theory is that iron was removed from the oceans by oxidation with oxygen [5]. The increase in sulfate levels may have been promoted by a rise in the atmospheric oxygen concentration and an increase in the oxidized sulfur reservoir which is seen as a natural extension of the continued oxidation of the Earth-surface environment [1].

The first evidence for oxygen production by oxygenic photosynthesis is found at around 2.8 Ga (**Figure 1**). Even so, the oxidation of the Earth-surface was quite protracted. As the waste of oxygenic photosynthesis, molecular oxygen initially reacted with sulfur, iron or methane but ultimately accumulated in the atmosphere. Sulfate did not accumulate in the oceans to concentrations >around 1 mM until about 2.3 Ga, roughly contemporaneous with other indicators of Earth-surface oxidation. In fact, around 2.4 billion years ago, in the Great Oxidation Event, atmospheric oxygen perhaps precipitated the first global ice age—a "snowball earth" [9]. Despite an apparently major accumulation of oxygen into the atmosphere during the early Proterozoic, various lines of biological and geological evidence suggest that oxygen levels did not surpass approx. 10% of present-day levels until much later in Earth history [10].

The evolution of oxygenic photosynthesis provided for dramatically increased rates of carbon production, and a much wider range of ecosystems for both carbon production and carbon oxidation. In oxygenic photosynthesis, electrons are transferred from water—split by chlorophyll photo-oxidized by the sun—via an electron-transport chain ultimately onto CO_2 to form sugars. In principle, by freeing photosynthesis from the availability of reduced chemical substances, the global production of organic carbon could be greatly increased. In this way, the evolution of oxygenic photosynthesis becomes a positive feedback on the oxidation of the Earth-surface environment as the ability to produce more carbon should cause more carbon burial, and this should lead to Earth-surface oxidation. Also of possible significance in promoting Earth-surface oxidation would have been the growth and stabilization of continental platforms where carbon burial could occur [11]. High carbon burial rates increased levels of atmospheric oxygen to >10% present-day levels, promoting the widespread oxidation of marine surface sediments. Present-day levels of atmospheric oxygen may not have been reached until the Neoproterozoic (0.54–1.0 Ga) in association with a second major burial episode of organic matter [12].

1.4. Prokaryotic evolution

Associated with volcanic activity was the introduction of inorganic compounds (H_2, CO_2, CH_4, SO_2, and H_2S) driving the metabolism of early chemolithoautotrophic organisms, including organisms of the sulfur cycle. Prokaryotic organisms of the domains archaea and bacteria were probably the only forms of life because they have the ability to metabolize sulfur compounds. The deepest-branching lineages within the bacteria house hyperthermophilic organisms; these include chemolithoautotrophic group, sulfate reducers genus, and the dominantly fermentative organisms [13].

The deep-branching organisms within the domain bacteria are hyperthermophilic. Branching a little farther up tree from the hyperthermophilic groups are the green non-sulfur bacteria which includes several anoxygenic phototrophs. An evidence for sulfate formation by anoxygenic phototrophic bacteria was provided in 3.4 Ga (**Figure 1**). These phototrophs are the deepest-branching photosynthetic organisms, and oxidize, via a single photosystem, hydrogen sulfide to elemental sulfur, and sulfate. Farther up are found the green sulfur bacteria, and beyond that, the tree indicates a tremendous radiation of bacterial life. Some of the more conspicuous groups include the cyanobacteria (oxygenic phototrophs), the purple bacteria (an enormous variety of heterotrophic, chemolithoautotrophic, and anoxygenic phototrophic bacterial types), and the gram-positive bacteria (including a wide variety of anoxygenic phototrophs, fermentative bacteria, and heterotrophic bacteria, including many thermophilic organisms). The domain bacteria house most of the prokaryotes with which we are most familiar. These organisms conduct an enormous range of metabolisms including fermentation, acetogenesis, sulfate reduction, elemental sulfur reduction, metal oxide reduction, denitrification, nitrification, aerobic respiration, oxygenic and anoxygenic photosynthesis, and the whole range of chemolithoautotrophic metabolisms using oxygen, nitrate, and possibly metal oxides as electron acceptors [1].

1.5. The role of oxygen

Oxygen was initially released as a by-product of photosynthesis following the emergence of blue-green algae. The evolution of oxygenic photosynthesis produced a dramatic increase in the primary production of organic material and promoted a profound expansion of the ecosystems available to prokaryotic life. Either associated with or following the evolution of oxygenic photosynthesis is the emergence of lineages housing most of the bacteria of which we are familiar, including most of the bacteria of the sulfur cycle. A dramatic evolution in bacterial life might have been made possible by: (1) the appearance of oxygen in sufficient quantities to fuel important metabolic pathways such as, for example, methanotrophy, and (2) an increase in available ecosystem space that global-ranging carbon production and subsequent carbon deposition onto sediments would provide [1]. An increase in oxygen levels would account for: (1) the initiation of large ^{34}S-depletions (>45 permil) in sedimentary sulfides, indicating the operation of the oxidative sulfur cycle; (2) the evolution of colorless sulfur bacteria; and (3) the emergence of large multicellular animals [10].

The evolution of oxygenic photosynthesis, like the evolution of life itself, was a singular event and not driven by any obvious environmental stimuli except for the opening of an enormous range of new environments that could support photosynthesis [1]. As atmospheric concentrations increased, it became possible to support more complex, multicellular life forms, including placental mammals [13]. Overall evolution for mammals living on dry land has been closely linked to adaptation to changes in the O_2 concentration in the environment [14, 15].

1.6. Phosphagenic energy

On Earth 4 billion years ago, there was virtually no oxygen and far higher levels of CO_2 (anything up to a thousand-fold more). When dissolved in water, CO_2 forms carbonic acid that is the difference CO_2 made and acidified the oceans. Back then, the pH of the oceans was likely to have been in the range of 5–6 [16]. Polyphosphates, such as ATP and pyrophosphate, form under acidic conditions at low water activity (in hydrophobic membranes) whereas their hydrolysis is favored under alkaline aqueous conditions [17]. As the alkaline fluids percolated into acidic oceans, through a labyrinth of interconnected micropores lined with hydrophobic iron-sulfur membranes, the vent system would have developed a natural proton gradient. It is therefore plausible that natural proton gradients could have driven the cycling between pyrophosphate and phosphate, or ATP and ADP, in the vent environment [18]. Whatever the mechanism, the first cells could not have left the vents without chemiosmotic coupling— nothing else could have provided the necessary energy. The vents also equipped the first cells with all the necessary tools—proton gradients, electron-conducting iron-sulfur clusters, and charged membranes. When the first prokaryotic cells did emerge, these were the tools of their trade. With them, they were set for photosynthesis. Thus, as a general rule, it is fair to say that prokaryotes can be classified not by their morphology but by their metabolic capabilities, and the most significant of those was photosynthesis [18]. Photosynthesis reverses respiration. Drawing on water as a fuel, rather than reduced chemicals derived from volcanic and hydro-thermal processes, probably increased global biomass 10-fold. Photosynthetic water-splitting transformed the planet. Additionally, photosynthesis in its oxygenic form changed the environmental O_2 concentration [9, 19].

The ancient methanogens (archaea) and acetogens (bacteria) presented probably the most ancient chemolithotrophic pathway in life, the direct reaction of hydrogen with carbon dioxide, known as the acetyl CoA pathway. This pathway provides both the carbon and energy metabolism of life—there is no need for solar power, primordial soup, ATP or any other accouterments [13]. In oxygenic photosynthesis, electrons are transferred from water—split by chlorophyll photo-oxidized by the sun—via an electron-transport chain that is exactly analogous to the respiratory chain, ultimately onto CO_2 to form sugars. The flow of electrons drives the transfer of protons across the thylakoid membranes to generate a proton gradient, which in turn drives ATP synthesis. Therefore, oxygenic photosynthesis is limited only by nutrient and light availability (water is nearly ubiquitous) and not by the availability of electron donors [20]. Oxygenic photosynthesis provided for diverse metabolic possibilities including nitrification, methanotrophy, and non-photosynthetic sulfide oxidation.

There are only six known pathways of carbon assimilation across all life, including the Calvin cycle (used in oxygenic photosynthesis), the reverse Krebs cycle (found in many vent bacteria), and the acetyl CoA pathway. All but the acetyl CoA pathway require an input of energy, in the form of ATP or some equivalent, which is provided by sunlight in photosynthesis and oxygen in the case of chemosynthesis in vents. In the case of the acetyl CoA pathway, 1 ATP must be spent to overcome the kinetic energy "hump"; but instead of reclaiming just one ATP, chemiosmotic coupling makes it possible to gain about 1.5 ATPs per CO_2. Therefore, only the acetyl CoA pathway, the direct reaction of H_2 with CO_2 can provide the energy required for growth in the absence of light or oxygen, and even this pathway can only do so by the way of chemiosmotic coupling. In chemiosmotic, the energy released by an exergonic reaction is used to transfer one or more protons across a membrane. So, as long as the energy released is sufficient to transfer a single proton at least part of the way across the membrane, the reaction can be repeated indefinitely to generate, in the end, a proton gradient. Gradient can be used independently to power ATP synthesis. In all forms of oxidative phosphorylation, the passage of electrons from the donor to the acceptor is coupled to ATP synthesis by the way of an intermediary proton gradient across a membrane—chemiosmotic coupling [13].

1.7. Respiration

Respiration, by necessity, evolved early. Respiration is typically divided into aerobic and anaerobic. Aerobic respiration obviously requires oxygen. Anaerobic respiration is typically taken to mean anaerobic glycolysis, or fermentation. The distinction is between substrate-level phosphorylations. In fermentation, the phosphate groups are transferred directly by chemistry. In oxidative phosphorylation, electrons are transferred from an electron donor such as glucose (but which could be other organic or inorganic donors such as Fe^{2+}) via a series of redox centers to a terminal acceptor. In aerobic respiration, this acceptor is oxygen. In anaerobic respiration, the electrons donated by anaerobic glycolysis or fermentation (probably hydrogen) is taken by a range of other electron acceptors (initially CO_2), from NO to Fe^{3+} to protons. On the early Earth, H_2S and Fe^{2+} would have been major electron donors and Fe-S clusters have the important intrinsic factor of transferring single electrons. Iron-sulfur clusters (Fe-S) are yet found at respiratory complexes: notably in I and II [13].

1.8. Evolutionary oxidative metabolism

Reactivity allows oxygen to participate in high-energy electron transfers, and hence support the generation of large amounts of adenosine-5-triphosphate (ATP) through oxidative phosphorylation. This is necessary to permit the evolution of complex multicellular organisms once O_2 enables more efficient energy transformation from dietary proteins, carbohydrates, and fats to ATP. ATP molecules provide the chemical energy required to conduct the biochemical reactions essential to cellular life including protein biosynthesis, active transport of molecules across cellular membranes, and muscular contraction [15].

1.9. Oxygen reactivity

Oxygen is hardly toxic if left to itself; but it is readily activated in the presence of the every respiratory chain that is necessary for life. The reactivity of oxygen, of course, is limited by kinetics. The kinetic limitation on the reactivity of oxygen relates to its unusual electron outer orbital structure, giving molecular oxygen two electrons in parallel spin. Most of the O_2 used during the oxidation of dietary organic molecules is converted into water via the combined action of the enzymes of the respiratory chain. Around 1–2% of the O_2 consumed escapes this process and is diverted into highly reactive O_2 free radicals and other reactive O_2 species (ROS) at a rate dependent on the prevailing O_2 tension. The term "reactive oxygen species" is applied to both free radicals and their non-radical intermediates. Free radicals are defined as species containing one or more unpaired electrons, and it is this incomplete electron shell that confers their high reactivity [13]. Free radicals can be generated from many elements, but in biological systems it is those involving oxygen and nitrogen that are the most important [21].

1.10. Superoxide production

Under normal conditions, 2% of oxygen consumed is converted to superoxide (SO) in the mitochondria rather than being reduced to water. Because of its charge, SO is membrane impermeable and so remains within the mitochondrial matrix [20]. Hence, under physiological conditions, the most common oxygen-free radical is the superoxide anion, and mitochondria are considered the principal source. Because the roots of respiratory chains are in geochemistry composition, notably the Fe-S clusters, all these respiratory chains become badly insulated wires in the presence of oxygen. Consequently, the transfer of electrons along the enzymes of the respiratory chain is not totally efficient, and the leakage of electrons on to molecular oxygen, in particular from complexes I and III, results in the formation of SO. ROS leak has more to do with the speed of electron flow down electron-transport chains than it does with the concentration of oxygen itself. In general, ROS leak is lower in state III respiration (when ATP consumption is fast) than it is in state IV respiration, when electron flow is limited by ADP deficiency [22]. If the respiratory complexes become highly reduced, they become more reactive with oxygen; and the higher membrane potential can drive electrons in reverse back into complex I, again increasing the rate of ROS leak.

The rate of superoxide formation is determined by the number of electrons present on the chain, and so is elevated under conditions of hyperoxia and of raised glucose. Paradoxically, it is also increased under conditions of hypoxia, when the reduced availability of oxygen that acts as the final electron acceptor for complex IV causes electrons to accumulate. Superoxide can also be generated through the leakage of electrons from the shorter electron-transport chain within the ER [23, 24]. About 25% of SO within cells is generated within the ER mostly by the formation of disulfide bonds during protein folding. This can increase in cells with a high secretory output, and also under conditions of ER stress when repeated attempts to refold misfolded proteins may take place. Other sources of superoxide under physiological conditions include the enzymes nicotinamide adenine dinucleotide phosphate (NADPH) oxidase, cytochrome P450, and other oxidoreductases. Hence, various growth factors, drugs, and toxins cause increased generation of ROS. Additionally, under pathological conditions, the enzyme xanthine dehydrogenase becomes an important contributor. This enzyme degrades purines, xanthine, and hypoxanthine to uric acid and, under normal conditions, uses NAD as the electron recipient. However, under hypoxic conditions, it is proteolytically cleaved to the oxidase form, which donates electrons to molecular oxygen. This enzyme plays a key role in the reperfusion phase of ischemia-reperfusion injury, when its action is augmented by the buildup of hypoxanthine as a result of ATP breakdown during the hypoxic period [25].

Superoxide is detoxified by the superoxide dismutase enzymes, which convert it to hydrogen peroxide. Two isoforms of superoxide dismutase convert SO to hydrogen peroxide, the manganese form that is restricted to the mitochondria and the copper and zinc form that is located in the cytosol [21].

1.11. Oxidant/anti-oxidant balance

Aerobic reactions lead to the accumulation of reactive oxygen species, which can be toxic to the cells. Therefore, oxygen has both positive benefits and potentially damaging side effects for biological systems. In fact, biotic and abiotic stresses can trigger a dramatic increase in the generation of reactive oxygen species such as superoxide radicals, hydroxyl radicals, and hydrogen peroxide in the intracellular environment. Consequently, our body is under constant oxidative attack from reactive oxygen species (ROS). Oxygen reactivity renders it liable to attack any biological molecule, be it a protein, lipid or DNA [21]. Hence, a complex system of antioxidant defenses has evolved that generally holds this attack in balance. In this context, aerobic organisms have developed several non-enzymatic and enzymatic systems to neutralize these compounds. The enzymatic systems include a set of gene products such as superoxide dismutases, catalases, ascorbate peroxidases and glutathione peroxidases (GPx). Enzymatic and non-enzymatic defenses inhibit oxidant attack. The enzymatic defenses all have a transition metal at their core, capable of taking on different valences as they transfer electrons during the detoxification process [26].

The concept of a pro-oxidant-antioxidant balance is central to an understanding of oxidative stress for several reasons. Firstly, it emphasizes that the disturbance may be caused through changes on either side of the equilibrium (e.g. abnormally high generation of ROS or deficiencies in the antioxidant defenses). Secondly, it highlights the homeostatic concentrations of ROS. The concept of a balance draws attention to the fact that there will be a graded response to oxidative stress. Hence, minor disturbances in the balance are likely to lead to homeostatic adaptations in response to changes in the immediate environment, whereas more major perturbations may lead to irreparable damage and cell death. The boundary between normal physiological changes and pathological insults is therefore inevitably indistinct. The definition of oxidative stress is necessarily broad because the outcome depends in part on the cellular compartment in which the ROS are generated. There are many potential sources of ROS, and the relative contributions of these will depend on the environmental circumstances prevailing. As the reactions of ROS are often diffusion-limited, the effects on cell function depend to a large extent on the biomolecules in the immediate vicinity. Different insults will therefore generate different outcomes [21].

1.12. Thiol redox

The term thiol refers to compounds containing sulfur. Sulfur-containing compounds are found in all body cells and are indispensable for life. Among plasma thiols, total Cys is the most abundant, followed by Hcy and GSH. Sulfur atoms are also important in the iron-containing flavoenzymes, such as, succinate dehydrogenase and NADH dehydrogenase [27]. The thiols are in a dynamic relationship through thiol-disulfide exchanges and redox reactions. Cys residues are susceptible to a variety of modifications by reactive oxygen and nitrogen oxide species (ROS and RNS). Oxidation by ROS or RNS can result in a disulfide bridge forming between two thiols, either within a protein chain or between protein chains [28]. The introduction of potential disulfide-forming thiol pairs may be facilitated by the fact that both Cys do not need to be introduced into the protein chain simultaneously. Acquisition of structural disulfides in proteins can potentially occur via transition through a redox-active disulfide state. Reaction of protein thiols with low-molecular weight thiols such as glutathione (GSH) can yield mixed disulfides. However, the formation of the disulfide-bonded form will only occur under conditions of oxidative stress. Hence, the incorporation of a single Cys may make the protein immediately responsive to a range of oxidative modifications. Cys can be nitrosated, glutathionylated, and can form covalent bonds with other Cys. RNS such as nitric oxide ($\bullet$NO) can mediate S-nitrosation to yield an S-nitrosothiol (RSNO). Other RNS, such as peroxynitrite (ONOO$-$), can also mediate S-nitration to yield S-nitrothiols (RSNO$_2$). Sequential oxidation of Cys thiols yields sulfenic ($-$SOH), sulfinic ($-$SO$_2$H), or sulfonic ($-$SO$_3$H) acid derivatives. Introduction of a second Cys at a later stage may then enable disulfide formation subject to further constraints [29].

1.13. Sulfur-containing antioxidants

Some of sulfur-containing antioxidant compounds are cysteine (Cys), methionine (Met), taurine (Tau), glutathione (GSH), lipoic acid, and mercaptopropionyl glycine. Glutathione

(L-gamma-glutamyl-L-cysteinylglycine) is the principal tripeptide thiol involved in the anti-oxidant cellular defense and a major hydro-soluble component of the cellular antioxidant system [30]. GSH is highly reactive and instills several vital roles within a cell including anti-oxidation, maintenance of the redox state, modulation of the immune response, and detoxification of xenobiotics. These reactions can be divided into those involved with the sulfhydryl moiety or with the gamma-glutamyl portion of the tripeptide [30]. In the former are included the oxidation-reduction reactions and the nucleophilic reactions in which the reduced sulfhydryl reacts with electrophiles to form a thioester [31].

In its antioxidant performance, the oxidation of the reduced form of glutathione (GSH) to form GSSG is carried out either by direct interaction with free radicals or, more often, when GSH acts as a cofactor for antioxidant enzymes such as GSH peroxidases. The activity of glutathione peroxidase depends on the presence of reduced glutathione (GSH) as a hydrogen donor [27]. Cytosolic GSH peroxidase reacts in peroxisomes with the hydrogen peroxide produced during the aerobic metabolism. In this reaction, GSH is oxidized to GSSG. In order to prevent oxidative damage, the GSSG is reduced to GSH by (riboflavin-dependent) GSSG reductase at the expense of NADPH (generated by pentose-shunt pathway), forming a redox cycle. Glucose-6-phosphate dehydrogenase is the first enzyme of the pentose-shunt pathway and this enzyme is subject to common polymorphisms, and decreased activity may compromise GSH concentrations. Other thiol compounds, such a thioredoxin, are capable of detoxifying hydrogen peroxide, but in turn require converting back to the reduced form by thioredoxin reductase [21, 27].

A function of GSH is the maintenance of the intracellular redox balance and the essential thiol status of proteins. In the reaction, the oxidized protein (protein-SSG) is reduced (protein-SH) and the reduced glutathione (GSH) is oxidized (GSSG). The equilibrium of this reaction depends on the concentrations of GSH and GSSG [25]. In extreme conditions of oxidative stress, the ability of the cell to reduce GSSG to GSH may be less, inducing the accumulation of GSSG within the cytosol. To avoid a shift in the redox equilibrium, the GSSG can be actively transported out of the cell or react with protein sulfhydryl groups and form mixed disulfides [32].

Introduction of a single Cys into a protein may allow reversible GSH conjugation to occur. Glutathione-cysteine adducts may be removed from proteins by glutaredoxin, whereas disulfides may be reduced by thioredoxin. Thiol compounds, such a thioredoxin, are capable of detoxifying hydrogen peroxide, but in turn require converting back to the reduced form by thioredoxin reductase [21]. Storage of Cys is another important function of GSH because Cys is extremely unstable extracellularly and rapidly auto oxidizes to cystine in a process that produces potentially toxic oxygen-free radicals [33]. The gamma-glutamyl cycle allows GSH to be the main source of Cys. In this cycle, GSH is released from the cell and the enzyme gamma-glutamyl transferase (yGT) transfers the y-glutamyl moiety of GSH to an amino acid (the best acceptor being Cys), forming y-glutamyl-amino acid and cysteinylglycine [32]. Cysteinyl-glycine is broken down by dipeptidase to generate Cys and Gly. Once inside the cell, the majority of Cys is incorporated into GSH, some being incorporated into protein and some degraded into sulfate and Tau. Similarly, the y-glutamyl-amino acid (Gln) can be transported back into the cell and once inside can be converted to Glu and used for GSH synthesis [32, 33].

2. Discussion

2.1. The role of oxygen and oxidative stress in prokaryotes evolution

Oxygen produces only about an order of magnitude more power than fermentation; and the difference between aerobic and true anaerobic respiration is somewhat less than that. While this is substantial, it is orders of magnitude less than the difference made by mitochondria, and probably differences in nutrient availability or concentration gradients outweighed any metabolic advantages of oxygen, at least among bacteria. Oxygen hardly wrought a global revolution in prokaryotic physiology. Even in the presence of oxygen, no prokaryote ever came close to evolving the morphological complexity of eukaryotes. In this context, the evolution of aerobic respiration may have made a difference, but the most immediate impact of the rising tide of oxygen was its juxtaposition with the electron-transport chains of bacteria, all of which transfer single electron [13]. The basic problem, which is central to eukaryotic evolution too, is that the rates of photooxidation and electron transfer, being essentially quantum events, differ from the rates of chemical reduction and carbon assimilation. This means that conditions such as high light intensity (which rapidly photo-oxidizes chlorophyll), low temperatures (electron transfers are barely slowed, but chemical reactions are much slower), and iron deficiency (leading to poor respiratory stoichiometry) all cause high ROS leak. Without compensation, then, ROS leak is largely defined by poor growth: by a low demand for ATP and highly reduced respiratory complexes. There are various ways out of this "high-voltage" situation, from mild uncoupling to complete depolarization of the membrane, or the use of alternative oxidases, which pass electrons directly on to oxygen, without coupling to proton translocation. All of them, in effect, short circuit the membrane potential, enabling faster electron flow, less reduced respiratory complexes and lower ROS leak. If this high ROS leak is not brought under control quickly, the caspase enzymes are activated, significantly by the loss of the respiratory carrier cytochrome c, in plants as well as animals, and the cell is eliminated. Much the same problems affect the respiratory chains of non-photosynthetic aerobic bacteria, among them the free-living ancestors of mitochondria, which likewise are capable of controlled cell death using metacaspase enzymes. The later development of apoptosis in metazoans makes use of enzymes that are bacterial in ancestry, notably the caspases, but also the Bcl-2 family and other mitochondrial apoptotic proteins [34]. Controlled cell death offers the advantage of recycling scarce nutrients, and so can be beneficial to the larger grouping, whether an organism, a colony, or selfish genes [13].

The point is that the evolution of metazoan cell death that require apoptosis was built on a system that evolved in relatively complex clonal bacteria capable of an apoptotic-style of cell death in response to oxidative stress. Consequently, the single greatest danger is the failure to pass electrons on swiftly down respiratory chains, resulting in highly reduced complexes in an aerobic atmosphere. The way in which these factors played out in the respiratory chains of eukaryotes may have been one of the most significant selective forces in eukaryote evolution [13]. However, there must be an adjustable threshold, above which ROS leak stimulates apoptosis and developmental failure, and below which ROS leak is tolerated (hormesis), or might even be beneficial as a redox signal [35]. A variable apoptotic threshold has profound implications for fertility, fecundity, adaptability, fitness, aging, and age-related disease. Setting the apoptotic threshold high, meaning a high tolerance of ROS leak before apoptosis is triggered, enables high fertility and fecundity. However, the offspring is less fit, and more likely to suffer from mitochondrial diseases. They will have lower aerobic capacity. Worst of all, they will leak ROS from their mitochondria at a faster rate, without triggering apoptosis. The outcome is a shorter lifespan, and a greater tendency to oxidative stress and chronic inflammatory conditions linked with aging, such as diabetes, cardiovascular disease, and cancer. In short, there is a trade-off between fertility, fecundity, and adaptability, on the one hand, and aerobic capacity, lifespan, and susceptibility to age-related disease on the other. The trade-off is mediated by sensitivity to oxidative stress [35]. Thus, oxygen introduces a new penalty for failure, controlled cell death, that later played a central role in the evolution of true multicellular organisms [13].

2.2. Redox signaling

Although ROS first came to the attention of biologists as potentially harmful by-products of aerobic metabolism, it is now recognized that they play important roles as secondary messengers in many intracellular signaling pathways [25]. This is because as hydrogen peroxide is nonpolar, it is able to diffuse through cell and organelle membranes, and hence acts widely as a second messenger in signal transduction pathways [21]. The redox system can modify functions of proteins through regulating their expression, post-translational modifications, and stabilities. Intracellular redox homeostasis regulates the expression of multiple gene-encoded proteins affecting cell death and survival. In response to alterations in oxidative status, the transcription of those genes can be modulated in part through a redox control of transcription factors such as NF-kB, AP-1, Nrf2, and HIF [36]. Upon exposure of cells to oxidative stress, signaling pathways such as protein kinase C, phosphatidylinositol-3 kinase, and MAP kinase, phosphorylate the transcription factor nuclear factor-erythroid 2-related factor 2 (Nrf2). After phosphorylation, Nrf2 translocates to the nucleus and binds to the antioxidant response element (ARE) within the promoters of genes encoding antioxidant enzymes and detoxifying enzymes. Key Nrf2 target genes include glutathione peroxidases (GPx), glutathione S-transferase (GST), superoxide dismutase (SOD), cytochrome P450, NAD(P)H quinone oxidoreductase, and heme oxygenase (HO) [27].

Thiol-based redox signaling is the collective name for biochemical pathways that regulate cellular processes by post-translational modification of sulfur moieties in cysteine (Cys) and methionine (Met) residues of proteins. A single Cys residue can form a redox-sensitive site on a protein. Thus, a redox-active disulfide may be introduced into a protein structure by stepwise mutation of two residues in the native sequence to Cys. By extension, evolutionary acquisition of structural disulfides in proteins can potentially occur via transition through a redox-active disulfide state. However, oxidation of a cytosolic molecule, promoting formation of the disulfide-bonded form, will only occur under conditions of oxidative stress [29].

When a single Cys is present in a protein, conjugation of the redox buffer glutathione may induce conformational changes, resulting in a simple redox switch that effects a signaling cascade. In its role as a redox buffer, GSH is conjugated to reactive Cys of endogenous proteins, inducing conformational changes in the substrate proteins, and effecting a signaling cascade that evokes biological responses [28].

A single Cys residue can form a potential redox-sensitive site on a protein because a second cysteine can be introduced into the sequence with a disulfide formation and oxidation of the cytosolic molecule. The formation of disulfide bridges between two Cys molecules is important in stabilizing protein conformation; therefore sulfur atoms in Cys are responsible for the major covalent cross-links in protein structures. Disulfide bonds between Cys residues are generally thought to confer extra rigidity and stability to their resident protein, forming a type of protein-aceous spot weld. Conformational changes are generally small, involving protein backbone, and are often accompanied by a local increase in protein disorder [28, 37]. Surface modification of proteins by GSH results in significant disorder of the GSH distal to the covalent bond [37].

GSH plays important roles in nutrient metabolism and regulation of cellular processes, including cell differentiation, proliferation, and apoptosis [38]. DNA-binding activity of transcription factors often involves critical Cys residues, and the maintenance of these residues in a reduced form, at least in the nuclear compartment, is necessary. The reversible thiolation of proteins is known to regulate several metabolic processes including enzyme activity, signal transduction, and gene expression through redox-sensitive nuclear transcription factors such as AP-1, NF-kB, and p53 protein [39]. GSH is involved in a variety of cell functions such as DNA repair, cell cycle, regulation of cell signaling, and transcription factors, GSH therefore can modulate the genes of cell proliferation, differentiation, and apoptosis. The molecular mechanism of how GSH modulates cell proliferation remains largely speculative. A key mechanism for GSH's role in DNA synthesis relates to the maintenance of reduced glutaredoxin or thioredoxin, which is required for the activity of ribonucleotide reductase, the rate-limiting enzyme in DNA synthesis [32].

Due to different roles of ROS in cell signaling and many human pathological processes, imbalance of GSH is observed in a wide range of pathologies including cancer, neurodegenerative disorders, cystic fibrosis, HIV, and aging [30, 40].

2.3. Modular redox switches in life evolution

It is generally believed that before multiple genome sequences were complete, the increased complexity of organisms correlated with the gene number. Hence, after completion of the first genomes, the small differences in gene number between simple unicellular eukaryotes and mammals forced revision of how complexity is encoded. Additional complexity at the organismal level is likely encoded at the molecular level by noncoding DNA [41]. Hence, increased complexity may also be encoded at the protein level. It is recognized that concatenation of existing domains through gene fusion, also known as protein domain mosaicism, encodes new functions in more complex organisms [42]. Studies on the changing amino acid content of proteins show that domains are also not static structures. Additional complexity added to protein domains in the form of redox and other switches likely increases the signaling capabilities of individual domains. In other words, nature is continually tinkering with these independent folding units: a domain from archaea may not have the same sophisticated set of switches as the homologous domain from a mammalian protein. Thus, two modes of acquisition of increased protein complexity have been demonstrated to date: protein domain mosaicism [42] and acquisition of allosteric control sites. Babu et al. [43] showed that Zn finger allosteric control sites are added to protein sequences via retrotransposons. Both are Cys-based sites and are known to be redox regulated.

3. Conclusion

The sulfur cycle participation in life evolution comes from ancient anoxygenic elemental sulfur reduction generating environmental sulfide incorporated as mitochondrial Fe-S for the electron-transport chains. Sulfur compounds that contributed for electron leakage and oxidative stress have counteractions by thiol participation either as antioxidant defensors and/or as redox-modulators or cell functions, influencing life evolution and contemporary diseases.

Acknowledgements

Special thanks to the Brazilian Research Funding CNPq (RCB researcher fellowship) and CAPES (HTK graduate fellowship).

Conflict of interest

The authors declare no conflict of interest.

Author details

Roberto C. Burini[1]*, Hugo T. Kano[2] and Yong-Ming Yu[3]

*Address all correspondence to: burini@fmb.unesp.br

1 Department of Public Health, Center for Nutrition and Exercise Metabolism, Botucatu Medical School, Sao Paulo State University, Botucatu, São Paulo, Brazil

2 Pathology Graduate Program, Sao Paulo State University, Botucatu, São Paulo, Brazil

3 Shriners Burns Hospital, Massachusetts General Hospital Harvard Medical School, Boston, MA, USA

References

[1] de C, Raiswell R. The evolution of the sulfur cycle. American Journal of Science. 1999; **299**:697-723

[2] Woese CR. Bacterial evolution. Microbiological Reviews. 1987;**51**:221-271

[3] Olsen GJ, Woese CR. Ribosomal RNA: A key to phylogeny. The FASEB Journal. 1993; **7**:113-123

[4] Whittaker RH, Margulis L. Protist classification and the kingdoms of organisms. Biosystems. 1978;**10**:3-18

[5]   Cloud PE Jr. A working model of the primitive earth. American Journal of Science. 1972;**272**:537-548

[6]   Collerson KD, Kamber BS. Evolution of the continents and the atmosphere inferred from Th-U-Nb systematics of the depleted mantle. Science. 1999;**283**:1519-1522

[7]   Taylor SR. The geochemical evolution of the continental crust. Reviews of Geophysics. 1995;**33**:241-265

[8]   Hattori K, Cameron EM. Archaean magmatic sulphate. Nature. 1986;**319**:45-47

[9]   Lane N. Oxygen: The Molecule that Made the World. OUP: Oxford, England; 2002

[10] Canfield DE, Teske A. Late Proterozoic rise in atmospheric oxygen concentration inferred from phylogenetic and sulphur-isotope studies. Nature. 1996;**382**:127-132

[11] Des Marais DJ. Isotopic evolution of the biogeochemical carbon cycle during the Proterozoic eon. Organic Geochemistry. 1997;**27**:185-193

[12] Kaufman AJ. An ice age in the tropics. Nature. 1997;**386**:227-228

[13] Lane N. The evolution of oxidative stress. In: Pantopoulos K, Schipper HM, editors. Principles of Free Radical Biomedicine. Vol. I. Chap. 1. Nova Sc. Publ. Inc.; 2011. ISBN: 978-1-61209-773-2

[14] Falkowski PG, Katz ME, Milligan AJ, Fennel K, Cramer BS, Aubry MP, Berner RA, Novacek MJ, Zapol WM. The rise of oxygen over the past 205 million years and the evolution of large placental mammals. Science. 2005;**2005**:2202-2204

[15] Burton GJ, Hempstock J, Jauniaux E. Oxygen, early embryonic metabolism and free radical-mediated embryopathies. Reproductive Biomedicine Online. 2003;**6**:84-96

[16] Russell MJ, Arndt NT. Geodynamic and metabolic cycles in the hadean. Biogeosciences. 2005;**2**:97-111

[17] Martin W, Russell MJ. On the origin of biochemistry at an alkaline hydrothermal vent. Philosophical Transactions of the Royal Society of London. Series B, Biological Sciences. 2007;**367**:1887-1925

[18] Lane N, Allen JF, Martin W. How did LUCA make a living? Chemiosmosis in the origin of life. BioEssays. 2010;**32**:271-280

[19] Canfield DE, Rosing MT, Bjerrum C. Early anaerobic metabolisms. Philosophical Transactions of the Royal Society B. 2006;**361**:819-1836

[20] Knoll AH, Bauld J. The evolution of ecological tolerance in prokaryotes. Earth and Environmental Science Transactions of The Royal Society of Edinburgh. 1989;**80**:209-223 [PubMed]

[21] Burton GJ, Jauniaux E. Oxidative stress. Best Practice & Research: Clinical Obstetrics & Gynaecology. 2011;**25**:287-299

[22] Barja G. Mitochondrial oxygen consumption and reactive oxygen species production are independently modulated: Implications for aging studies. Rejuvenation Research. 2007;**10**:215-224

[23] Cadenas E, Davies KJA. Mitochondrial free radical generation, oxidative stress, and aging. Free Radical Biology and Medicine. 2000;**29**:222-230

[24] Tu BP, Weissman JS. Oxidative protein folding in eukaryotes: Mechanisms and consequences. The Journal of Cell Biology. 2004;**164**:341-346

[25] Droge W. Free radicals in the physiological control of cell function. Physiological Reviews. 2002;**82**:47-95

[26] Fink RC, Scandalios JG. Molecular evolution and structure-function relationships of the superoxide dismutase gene families in angiosperms and their relationship to other eukaryotic and prokaryotic superoxide dismutases. Archives of Biochemistry and Biophysics. 2002;**399**:19-36

[27] Burini RC, Lamônica VC, Moreto F, Yu YM, Henry MACA. Thiol metabolic changes induced by oxidative stress and possible role of B-vitamins supplements in esophageal cancer patients. Chapter 5. In: Glutathione. ISBN: 978-1-63463-372-7. Wilber © 2015 Nova Science Publishers, Inc

[28] Wouters MA, Iismaa S, Fan SW, Haworth NL. Thiol-based redox signalling: Rust never sleeps. The International Journal of Biochemistry & Cell Biology. 2011;**43**:1079-1085. DOI: 10.1016/j.biocel.2011.04.002

[29] Mohanasundaram KA, Haworth NL, Grover MP, Crowley TM, Goscinsk A, Wouters MA. Potential role of glutathione in evolution of thiol-based redox signaling sites in proteins. Frontiers in Pharmacology. 2015;**6**:10. DOI: 10.3389/fphar.2015.00001

[30] Balendiran GK, Dabur R, Fraser D. The role of glutathione in cancer. Cell Biochemistry and Function. 2004;**22**(6):343-352. DOI: 10.1002/cb0f.1149

[31] Tamba M, Quintiliani M. Kinetic studies of reactions involved in hydrogen transfer from glutathione to carbohydrate radicals. Radiation Physics and Chemistry. 1984;**23**:5

[32] Traverso N, Ricciarelli R, Nitti M, Marengo B, Furfaro AL, Pronzato MA, et al. Role of glutathione in cancer progression and chemoresistance. Oxidative Medicine and Cellular Longevity. 2013;**2013**:972913. DOI: 10.1155/2013/972913

[33] Burini RC, Moreto F, Borges-Santos MD, Yu YM. Plasma homocysteine and thiol redox states in HIV(+) patients. In: McCully KS, editor. Homocysteine: Biosynthesis and Health Implications. Nova Science Publishers; 2013. p. 14

[34] Koonin EV, Aravind L. Origin and evolution of eukaryotic apoptosis: The bacterial connection. Cell Death & Differentiation. 2002;**9**:394-404

[35] Lane N. On the origin of barcodes. Nature. 2009;**462**:272-274

[36] Trachootham D, Lu W, Ma O, Rivera-Del Valle N, Huang P. Redox regulation of cell survival. In: Sadoshima J, Sen CK, Tang B-L, Webster K, editors. Antioxidants & Redox Signaling. Vol. 10(8). Mary Ann Liebert, Inc; 2008. DOI: 10.1089/ars.2007.1957

[37] Mallis RJ, Poland BW, Chatterjee TK, Fisher RA, Darmawan S, Honzatko RB, et al. Crystal structure of S-glutathiolated carbonic anhydrase III. FEBS Letters. 2000;**482**:237-241. DOI: 10.1016/S0014-5793(00)02022-6

[38] Abdalla MY. Glutathione as potential target for cancer therapy; more or less is good? Jordan Journal of Biological Sciences. 2011;**4**(3):6

[39] Townsend DM, Tew KD, Tapiero H. The importance of glutathione in human disease. Biomedicine & Pharmacotherapy. 2003;**57**(3-4):145-155

[40] Borges-Santos MD, Moreto F, Pereira PC, Ming-Yu Y, Burini RC. Plasma glutathione of HIV(+) patients responded positively and differently to dietary supplementation with cysteine or glutamine. Nutrition. 2012;**28**(7-8):753-756. DOI: 10.1016/j.nut.2011.10.014

[41] Mattick JS. RNA regulation: A new genetics? Nature Reviews. Genetics. 2004;**5**:316-323. DOI: 10.1038/nrg1321

[42] Patthy L. Evolution of the proteases of blood-coagulation and fibrinolysis by assembly from modules. Cell. 1985;**41**:657-663. DOI: 10.1016/s0092-8674(85)80046-5

[43] Babu MM, Iyer LM, Balaji S, Aravind L. The natural history of the WRKY–GCM1 zinc fingers and the relationship between transcription factors and transposons. Nucleic Acids Research. 2006;**34**:6505-6520. DOI: 10.1093/nar/gkl888

Chapter 3

The Important Functions of GSH-Dependent Enzyme Glutaredoxin 2 (Grx2)

Christy Xavier, Xiaobin Liu, Yang Liu and Hongli Wu

Additional information is available at the end of the chapter

http://dx.doi.org/10.5772/intechopen.78653

Abstract

Reactive oxygen species (ROS) are generated at a very high rate throughout our lives as part of normal aerobic life. Glutathione (GSH), normally an antioxidant molecule that scavenges free radicals, oxidizes to form glutathione mixed disulfide (GSSG). As the GSSG/GSH ratio increases, GSSG naturally adds to other proteins, causing protein glutathionylation. Protein glutathionylation, defined as the reversible formation of a mixed disulfide (PSSG) between protein thiols (P-SH) and glutathione (GSH), appears to be the most important mode of thiol oxidation. In my chapter, we will discuss the important roles of GSH and GSH-dependent enzymes in health and disease, with the emphasis on glutaredoxin and thioredoxin systems. Their structures, catalytic reaction mechanisms, major physiological functions, and associations with diseases will be summarized in my chapter. We will also mention how GSH-dependent enzymes play a role in each major organ systems including the nervous, cardiovascular, immune, and visual system.

Keywords: glutathione (GSH), glutaredoxin (Grx), thioredoxin (Trx), the nervous system, cardiovascular system, immune system, visual system

1. Oxidative stress and protein glutathionylation

Reactive oxygen species (ROS) are generated at a very high rate throughout our lives as part of normal aerobic life. In particular, the mitochondria due to its high metabolic capacity are the primary site of endogenous ROS production. With each run of the mitochondrial electron transport chain, 1–2% of free radicals are predicted to escape the mitochondria [1]. With time,

ROS accumulation can be detrimental, as they can cause oxidative damage to proteins, lipids, DNA, and other crucial biological molecules [2]. Among these macromolecules, proteins are very sensitive to oxidative modification. Cysteine residues are particularly reactive with ROS due to the presence of thiol (-SH) group, which can be oxidized to sulfenic (SOH), sulfinic (SO$_2$H), sulfonic acids (SO$_3$H), or formed disulfide bonds (S-S). Protein glutathionylation, defined as the reversible formation of a mixed disulfide (PSSG) between protein thiols (P-SH) and glutathione (GSH), appears to be the most important mode of thiol oxidation. GSH, normally an antioxidant molecule that scavenges free radicals, oxidizes to form glutathione mixed disulfide (GSSG). As the GSSG/GSH ratio increases, GSSG naturally adds to other proteins, causing protein glutathionylation. In most cases, the addition of GSSG to proteins renders them inactive, but for specific proteins such as heat shock proteins or peroxiredoxin 6, it may serve as an activation mechanism or a cell signaling event [3, 4]. In small amounts, PSSG may alert the cell that oxidative stress is present and may lead to certain signaling cascades to restore a healthy cellular redox state. Nonetheless, excessive PSSG can be lethal because of severe protein inactivation and damage [5]. Therefore, maintenance of a redox state can translate to increased cell survival and protection.

2. Introduction to the glutaredoxin (Grx) system

To combat oxidative stress, the body is equipped with several antioxidant enzymes in order to restore the redox balance and protect the cell. Several antioxidant enzymes such as catalase, superoxide dismutase (SOD), and thioredoxin (Trx) have been thoroughly researched due to their effectiveness and ability to directly target and scavenge ROS [2, 6–8]. However, as it is becoming more clear that PSSG is an important post-translational modification linked to oxidative stress, recent research has highlighted the glutaredoxin (Grx) system, an antioxidant system capable of reversing PSSG formation. As shown in **Figure 1**, the Grx system

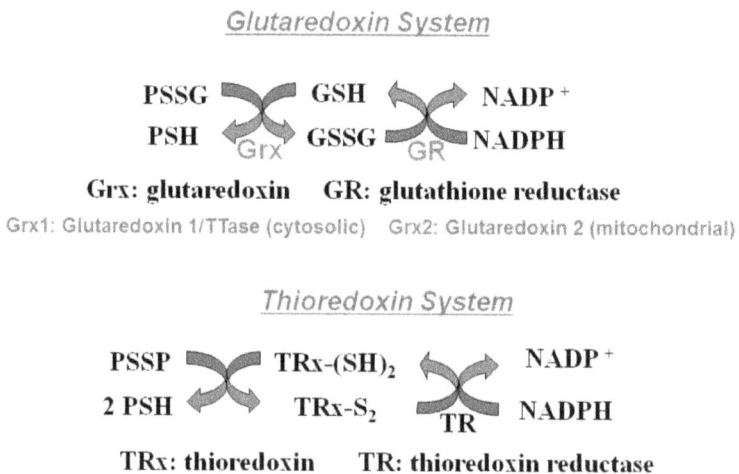

Figure 1. Glutaredoxin and thioredoxin system.

has two main subsets in humans: the cytosolic glutaredoxin 1 (Grx1) and the mitochondrial glutaredoxin 2 (Grx2) [7]. Grx1 has garnered much interest due to its similarity with Trx1 in promoting cytosolic protection, yet Grx2 with its primary mitochondrial localization may hold a crucial role in preventing cell death. This is especially important considering that mitochondria are the primary sites of ATP and ROS production and are critically involved in balancing pro- and anti-apoptotic signals [1, 4]. However, to date, only a few studies have been published to highlight Grx2's potential roles in humans. This chapter hopes to emphasize Grx2 as an imperative and crucial antioxidant enzyme, encourage more research studying Grx2's possible roles and protein targets, and provide support for Grx2 activating drugs to treat oxidative stress-induced diseases.

3. Glutaredoxin 2 (Grx2)

3.1. Grx2 localization and expression

Grx2 was first discovered simultaneously by two research groups leading by Holmgren and Gladeshev in the same year 2001. Both teams characterized Grx2 as a 18 kDa oxidoreductase enzymes capable of reversing PSSG and working as electron donors for ribonucleotide reductase [9]. There are several alternative spliced forms in different types of organisms, but in humans specifically, there are two main subsets: the cytosolic Grx1 and the mitochondrial Grx2. Grx1 has also been implicated in the mitochondrial inner membrane space but in miniscule amounts. Grx2 is rather unique, in that depending on cell type and organism, may be present in different cellular components. Grx2 has three alternatively spliced forms: Grx2a, Grx2b, and Grx2c. Grx2a is the mitochondrial isoform, whereas Grx2b and Grx2c mainly reside in the cytoplasm and nucleus. Grx2a is ubiquitous in every cell, except in the testes [6, 10, 11]. In spermatids, Grx2c is prominently present in the cytoplasm, yet Grx2a is absent (found to be less than 1% in the mice testes) [12, 13]. In embryos, Grx2 expression tends to surpass other dominant antioxidant enzymes such as Trx1, Trx2, and Grx1. During the gestation of E11 embryos, more than 50% of Grx2 mRNA transcripts compose of Grx2a [13]. This is not remarkable when considering that aerobic respiration and mitochondria formation takes precedent in this stage of gestation. Surprisingly, Grx2b and Grx2c are rarely found in normal human cells, but its presence has been discovered in HeLa cells and certain cancer cell lines. In patients with hepatocellular carcinoma and underlying metabolic syndrome that increases a person's risk for diabetes, stroke, and heart disease, Grx2 expression was particularly high [14]. In non-small cell lung cancer and adenocarcinoma, proliferative effects are correlated with higher cytoplasmic and nuclear Grx2 levels [15]. Along with several studies supporting Grx2 is highly expressed in cancer cells, Grx2 may have a distinct and special function in cancer cells, possibly in enhancing cancer cell survival and proliferation [15–17].

Patterns of Grx2 expression vary from species to species. Unsurprisingly, Grx2's localization in mice showed that Grx2a was present in all tissue types. However, it was also identified that Grx2b and Grx2c mRNA transcripts were found at various expression levels in several tissues with higher concentrations in the heart, liver, kidney, and eyes [12, 13]. Another study looking at porcine ocular tissues showed that Grx2 is abundantly present in the ciliary body and retina but is lacking in the vitreous humor [18]. Both the ciliary body and retina have direct access to blood vessels and are also known to be more sensitive to oxidative stress, increasing the risk in these areas for a variety of oxidative stress-induced eye diseases. Therefore, Grx2's concentration in these areas indicate Grx2's potential role in preventing oxidative-induced damage in these exposed tissues.

By using artificial cloning of recombinant human Grx2 (hGrx2) in E. coli and HeLa cells, Grx2 was able to be fully expressed in the mitochondria and nucleus. Closer analysis suggests that Grx2a has a distinctive mitochondrial N-terminal signal sequence, yet Grx2a and Grx2b both have an arginine and lysine rich C-terminal sequence that highly resembles a nuclear translocation sequence [11]. Staining also suggested that Grx2a was found exclusively in the mitochondria, whereas Grx2b had a strong presence in the cytoplasm and a weaker presence in the nucleus. This contrasted to data previously shown that in Jurkat cells, Grx2 preferentially is found in the nucleus compared to the mitochondria [11]. Nonetheless, the reasons for these differences are unclear, other than the obvious alternative splicing of Grx2 transcripts and the absence of a translocation signal in Grx1 transcripts. Moreover, what prompts the predilection of Grx2a in the mitochondria despite having a nuclear translocation signal and the role of Grx2b and Grx2c in human cells still remain unsolved mysteries.

3.2. Grx2 structure

Compared to the exclusively monomeric and constitutively expressed Grx1, Grx2 is normally 18 kDa and found in an inactive dimeric form with an iron-sulfur cluster [19, 20]. As shown in **Figure 2**, the iron-sulfur cluster is crucial for Grx2's typical functioning, as it is able to detect oxidative stress in the environment. In terms of voltage differences, one study claimed that Grx2's dimeric form only disassociated with pulses greater than 0.5 V [21]. When oxidative stress is present, Grx2 cleaves to its active 16 kDa monomeric form and directly scavenges free radicals and reverses PSSG. Of all the isoforms, Grx2b does not have the specific iron–sulfur cluster and remains a monomer in non-stressed and stressed conditions [10, 21].

3.3. Grx2 mechanism

Grx1 and Grx2 have a particular active site template: Cys-X-Tyr-Cys. Grx1 has a proline residue after the first cysteine residue, whereas Grx2 has a serine instead. Although this gives Grx2 a slightly slower reaction rate, the serine essentially gives Grx2 a higher affinity for glutathionylated substrates [22]. Both monothiol and dithiol mechanisms can be performed, yet the monothiol mechanism using the N-terminal Cys is more accepted. Both Grx1 and

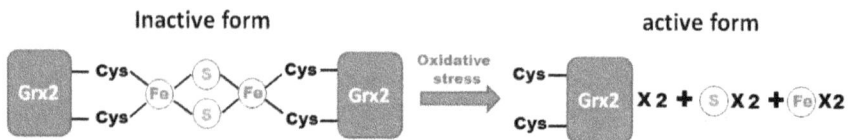

Figure 2. Glutaredoxin 2 as an iron-sulfur (Fe-S) protein.

Grx2 employ the same mechanism to remove GSH from its protein targets: it accepts the GSH from the protein adduct and becomes Grx-SSG only to be returned back to its reduced form (Grx-SH) by glutaredoxin reductase (GR) and the reducing power of NADPH. GSSG is further processed and recycled to GSH by the enzyme glutathione reductase [23, 24].

The cysteines in the active site of Grx2 play a major role for Grx2 to function. Along with active site Cys residues, there are several disulfide bridges between non-active site Cys residues that promote further stability in Grx2 structure [25]. This attribute has been remarkably the only conserved feature found in all vertebrate-specific Grx2 forms, indicating the necessity of this disulfide bridge in promoting Grx2's potency. However, other modifications of Cys residues may inhibit Grx activity. A study performed by Hashemy et al. indicated that compared to its counterpart Grx1, Grx2 enzyme activity was relatively initiated by S-nitrosylation, leading to the conversion of Grx2 into its active monomeric form [17]. Grx1's cysteine oxidation either by S-nitrosylation or intradisulfide bonds only lead to significant structural changes and inactivation. Therefore, compared to Grx1, Grx2 is more robust in highly oxidative stressed environments and is still able to function despite Cys modifications [17]. Moreover, this evidence highlights the ability of Grx2 to be exclusively activated under oxidative stress conditions.

4. Grx2 is a potent antioxidant and anti-apoptotic enzyme in different systems

4.1. The nervous system

Numerous research studies have defined Grx2 as a potent antioxidant and anti-apoptotic enzyme in the brain. Compared to other tissues, the brain is more vulnerable and sensitive to oxidative stress. Moreover, mitochondrial dysfunction is a similar theme found in several neurodegenerative diseases, including Parkinson's, amyotrophic lateral sclerosis (ALS), and Alzheimer's disease [26]. Because mitochondria are heavily involved in the apoptotic pathway, the ability to prevent mitochondrial induction of cell death and repair oxidatively damaged neurons are methods believed to be the future treatments for neurodegenerative disorders.

In motor neurons, aberrant SOD1, misfolded protein aggregation, oxidative stress, and mitochondrial dysfunction are all key players in ALS pathogenesis [27]. Ferri et al. was able to determine that Grx1 overexpression may increase mutant SOD1 solubility in the cytosol but has no effect in its solubility in the mitochondria or motor neuron apoptosis [27]. The accumulation of mutant SOD1 is attributed to higher rates of apoptosis due to ATP and GSH depletion and unusual mitophagy. A previous study in yeast has shown that GSH is a main contributor for SOD1 activation and glutathionylation especially when using oxidative stress agents like menadione are used and is crucial for lifespan development [28]. Rather, Grx2 overexpression cleared mutant SOD1 in the mitochondria, resulting in decreased mitochondrial fragmentation and abolishing neuron apoptosis [27]. Optic atrophy 1 (OPA1), a pro-fusion protein, and dynamin-related protein 1 (DRP1), a mitochondrial fragmentation protein, are often imbalanced due to mutant SOD1. However, Grx2 overexpression leads to the phosphorylation and activation of DRP1, restoring mitochondrial morphology and the mitochondrial GSH pool [27]. Overall, Grx2 was able to better reduce mutant SOD1 toxicity and increase neuronal cell survival, perhaps by working to correct oxidized disulfide bonds found in aggregated mutant SOD1.

The role of Grx2 in preventing Parkinson's pathology is also a recent research interest. The reduction of mitochondrial GSH pools may indirectly contribute to Grx2 impairment and the development of Parkinson's disease. Several studies have suggested that in dopaminergic neurons of the substantia nigra, the high GSSG/GSH ratio leads to Grx2 inhibition, causing a mitochondrial iron overload that deters the functions of complex I and aconitase, a tricarboxylic acid cycle (TCA) enzyme involved in converting citrate to isocitrate [29]. As a result, mitochondrial dysfunction due to reduced mitochondrial biogenesis and iron–sulfur cluster regeneration is rampant. When 1-methyl-4-phenyl-1,2,3,6,tetrahydropyridine (MPTP) is used to induce neurodegeneration in the extrapyramidal system, Grx2 responded with an increase in mRNA and protein levels after 4 h of treatment and helped alleviate MPTP toxicity and apoptosis [30]. Correspondingly, inhibition of Grx2 lead to compromised complex I activity, indicating Grx2's potential role in restoring complex I activity in times of oxidative stress [30, 31]. Another oxidative stress agent, 6-hydroxy-dopamine (6-OHDA), used to stimulate a Parkinson's disease model in neuroblastoma cell line, and *C. elegans* was found to increase the expression of Trx1, Trx2, Grx1, and Grx2. Grx2 was found to reduce 6-OHDA and thus prevent its cytotoxicity in dopaminergic neurons [29].

In an ischemia/reperfusion (I/R) model of the rat brain, Grx2 was the most affected, having rebounded from low levels during hypoxia to normal levels by reoxygenation [32]. When Grx2 was successively silenced, neuronal integrity was severely impacted, as there were more distinct areas of neuron damage and death. Furthermore, considering the use of perinatal rat brains, Grx2 was vital in the development of a normal neuronal phenotype [32].

Previous studies have also supported that Grx2 is fundamental for neuronal development. In a zebrafish model, Grx2c was found to be important for the development of the axonal scaffold, as silencing of Grx2c ultimately prevented the neuron differentiation and promoted neuron cell death in nearly 97% of embryos [33]. In particular, Grx2 mediated the redox regulation of collapsin response mediator protein 2 (CRMP2/*DPYSL2*), a protein that mediates axonal outgrowth, cytoskeleton remodeling, and neuronal cell migration [34]. However, their study also found that knockdown of Grx2 did not particularly induce carbonyl oxidative stress in the embryos. A follow-up study characterized the Cys-504-Cys-504 dithiol-disulfide switch as the regulatory mechanism for CRMP2 to control axonal development and neuronal differentiation [32]. Grx2 can reduce the two cysteines, causing CRMP2 conformational changes and activation for proper neuronal development [33, 34].

4.2. Cardiovascular and immune systems

Cardiac cells depend on mitochondrial metabolism for the constant ATP supply necessary to pump blood throughout the body. Compared to the other body organs, the heart has the most number of mitochondria per cell with nearly 5000 mitochondria per cardiac muscle cell [38]. Considering the importance of mitochondria in cardiac function, it is hypothesized that Grx2 plays a major role in protecting the heart's mitochondria.

Inhibition of Grx2 in zebrafish embryos' hearts prevent cardiac neural cells from entering the primary heart region [39]. This results in obstructed blood flow to the aorta and common cardinal vein. Proper heart looping and cardiac neural crest cell migration were also hindered by Grx2 knockdown. Because migration is highly dependent on actin polymerization, Grx2 knockdown would result in more globular actin (G-actin) production and prevent appropriate cell migration [37, 40]. Grx2 has also been attributed to vascular development in embryos. Sirtuin 1 (SIRT1) is prone to redox regulation with approximately 17 cysteines and 5 cysteines specifically known to be glutathionylated. Conserved Cys204 in the catalytic site is around the NAD+ binding site. NAD+ is a crucial cofactor that allows for SIRT1 functioning. Glutathionylation of the key Cys residue would block the cofactor binding site, leading to decreased SIRT1 activity [41]. Therefore, inhibition of Grx2 would promote glutathionylation of the active site, causing SIRT1 inactivation and prevent angiogenesis [41].

Grx2 gene Knockout (KO) mice have decreased aerobic respiration and ATP production due to complex I glutathionylation and inhibition [39]. As a side effect, hypertension ensued with increased cardiac glucose uptake. KO hearts also especially developed left ventricular hypertrophy and fibrosis, yet cardiac contraction and relaxation remained the same between wildtype and KO mice [39, 42].

Usually, with aging, there is a decline in antioxidant expression and activity. Interestingly, Gao et al. first suggested that Grx2 is found to be increased nearly twofold in the mitochondrial matrix of cardiac muscle cells [43]. However, most of the Grx2 is prominently in the inactive dimer formation. Only by exogenous superoxide production by xanthine oxidase was the dimer able to cleave to monomeric form [43]. H_2O_2 had no effect on Grx2 activation. When complex I inhibitor was used to induce endogenous ROS, there was not enough of an ROS threshold to activate Grx2 [41, 43].

Although there is no direct evidence hinting on the connection between Grx2 and immune cells, macrophages depend heavily on actin polymerization to migrate to areas of injury and inflammation. Actin has two prominent forms: filamentous F-actin and globular G-actin. G-actin is often glutathionylated, which prevents its conversion to F-actin [44]. Therefore, Grx2 activation can lead to increased F-actin production, which can cause increased cell migration, differentiation, and proliferation [24, 37, 44].

4.3. Skeletal muscle and adipose tissue

Research is still relatively scarce on how Grx2 may affect the cell bioenergetics, despite the consensus being that Grx2's regulation of complex I glutathionylation is a huge determinant on total ATP production.

In skeletal muscle, the GSH/GSSG mitochondrial pool is relatively high to account for high respiratory rates, which may be a consequence of having somewhat high levels of Grx2 and Trx2 in the mitochondria. Uncoupling protein 3 (UCP3) reversible glutathionylation is carefully mediated by Grx2 and plays a crucial role in promoting glucose metabolism and controlling oxidative stress in skeletal muscle [45, 46]. On the other hand, brown adipose tissue (BAT) is specifically saturated with mitochondria rich in iron, giving BAT its unique color and is highly concentrated in GSSG rather than GSH. Grx2 is lacking in BAT mitochondria, whereas Trx2 has slightly greater expression to most likely compensate for the lack of Grx2 [45]. Uncoupling protein 1 (UCP1) is similarly regulated by glutathionylation for thermogenesis, but the smaller need for immediate regulation prompts Grx1 and Trx2 to perform UCP1 reduction, albeit at a much lower efficiency [45]. Differences in the choice of substrates, demand, and rate for metabolism in stages of fasting and exercise could explain the lesser role glutathionylation has on BAT mitochondria bioenergetics. Moreover, to maintain a steadfast high iron environment in mitochondria, Grx2 may have to be suppressed, considering Grx2's distinctive iron-sulfur core that differs from the rest of the thioredoxin superfamily. Also, an oxidizing GSSG pool is preferred in BAT mitochondria, and Grx2 expression could alter bioenergetics more to a reducing GSH pool and hinder BAT's ROS production to generate heat. These are all possibilities that could explain why Grx1 and Trx2 are favorably expressed in BAT mitochondria unlike Grx2's key localization in skeletal muscle mitochondria. The selectivity of cysteine glutathionylation in UCP1 and UCP3 provides a great insight into skeletal muscle and BAT functioning [45–47]. The mechanism or the relative importance of Grx2 in regulating these functions remains unseen but does highlight Grx2's potential role in curbing obesity and insulin resistance in diabetes.

4.4. Cancer

Cancer cells are in a constant oxidative stress environment. In order to combat excessive oxidative stress-induced cell death and increase survival, cancer cells often overexpress antioxidant enzymes [2, 14, 15]. With higher expression of antioxidant enzymes, there is increased drug resistance to anti-cancer drugs such as doxorubicin whose primary purpose is to increase oxidative stress and create DNA damage to cause cancer cell death. Several previous studies have highlighted that hepatocellular carcinoma and adenocarcinoma tend to have higher amounts of Grx2 compared to healthy tissue [14, 15]. The role of Grx2 in cytoprotection and promoting drug resistance is still uncertain, but some studies have attempted to decipher how the knockdown or inhibition of Grx2 would affect cancer cell survival.

Enoksson et al. overexpressed Grx2 in HeLa cells, which caused less sensitivity to doxorubicin and 2-deoxy-D-glucose (2-DG). Grx2 was found to inhibit cytochrome c release [48]. Cardiolipin binds to cytochrome c to the inner mitochondrial membrane as an anti-apoptotic measure. When cardiolipin is reduced or lost, cytochrome c is free to move to the cytoplasmic membrane and initiate apoptosis. Without cytochrome c release, caspase is not activated, and apoptosis is thwarted [49]. It is believed that Grx2 decreases cytochrome c release by enhancing cardiolipin expression through prevention of cardiolipin glutathionylation [48, 49].

Similarly, Lillig et al. discovered that siRNA-mediated silencing of Grx2 sensitized HeLa cells to doxorubicin and phenylarsine oxide (PAO) severely. For doxorubicin, the ED50 dropped from 40 to 0.7 μM, whereas PAO's ED50 dropped from 200 to 5 nM [50]. When treated with a Grx1 inhibitor, HeLa sensitivity remained the same. Carbonyl stress was also slightly increased with Grx2 inhibition and doxorubicin treatment. This indicates that Grx2 potentially enhances drug resistance by altering the redox state of the HeLa cell. Conversely, it was also shown that Grx2 overexpression attenuated apoptosis by enhancing mitochondrial protection [48, 50].

In a renal I/R model, Grx2's increased expression in proximal tubular cells corresponded with less oxidative damage and injury after hypoxic conditions. Moreover, there was increased cellular survival and proliferation in HEK293 and HeLa cells after reoxygenation [16]. Considering Grx2's optimal localization in the kidney tubules, this may explain regeneration and anti-apoptotic effects even after an ischemic attack.

4.5. Cataract

Compared with other tissues, the ocular tissues are more susceptible to oxidative stress due to daily exposure of sunlight and relatively high oxygen consumption. Many major blind-inducing diseases, like cataract, age-related macular degeneration (AMD), glaucoma, and diabetic retinopathy are all proved to be closely related with oxidative stress. Therefore, anti-oxidant enzymes play a fundamental role in regulating redox homeostasis and maintaining overall ocular health. Lou's group is the first research team to investigate Grx2 distribution and compare its enzyme activity in ocular tissues. Their data have shown that Grx2 is present in all ocular tissues except vitreous humor. Grx2 level was highest in the mitochondrial-rich and vascular-rich tissues such as the retina while lowest in the tissues that are non-vascular or low mitochondria presence, such as the lens [18]. Fernando et al. first explored the function of Grx2 in lens epithelial cells and determined that human Grx2 has direct peroxidase activity and is capable of using GSH and thioredoxin reductase (TR) as electron donors [35]. Mouse Grx2 did not have as much of a potent peroxidase activity. Moreover, Grx2 could directly detoxify H_2O_2, tert-butyl hydroperoxide (tBHP), and cumene hydroperoxide with greater affinity for tBHP and cumene hydroperoxide [23, 35]. *Grx2* KO in lens epithelial cells also has been shown to less cell viability, decreased ATP levels, and increased complex I inactivation and glutathionylation when treated with H_2O_2 [36].

The main function of the lens is to maintain transparency so that the light can be transmitted and properly focused on the retina for vision. The eye lens contains high concentration GSH as the first line of defense against free radicals to protect critical enzymes and structural proteins from oxidation. However, with aging, the GSH pool decreases and oxidized GSH (GSSG) may attack protein thiols and formed PSSG. Many of lens structural proteins contain high level of thiol residues which are highly susceptible to oxidative damage. For example, alpha A-crystallin is one of the important and abundant structural proteins in the lens. It functions as a chaperone to facilitate protein folding and stabilize other lens proteins in a soluble form. However, under oxidative stress, many of its -SH groups could be gluta-thionylated, resulting in its chaperonelike function to be drastically lost. By using shotgun proteomic analysis, Giblin's group studied the glutathionylation sites of crystallins in the lens of the guinea pig under oxidative stress. Their data showed that almost all major crystallins, except αβ, were glutathionylated after the oxidative insult. More than 70% of the –SH groups were conjugated with GSH, cysteine, or both molecules. Interestingly, our previous data have shown that mice lacking Grx2 showed faster progression of cataract during aging. In the Grx2 null mice, the lens nuclear opacity began at 5 months, 3 months sooner than that of the control mice and the progression of cataract were also much faster than the age-matched controls. Importantly, this early cataract formation is closely associated with alpha A-crystallin glutathionylation accumulation and compromised mitochondrial functions in

Grx2 gene KO mice [37]. These data suggested that oxidative stress-induced thiol/disulfide imbalance at the stage of PSSG accumulation is the first critical checkpoint to prevent the cascading events leading to cataract formation. The findings also suggest that Grx2 protein may be a potential anticataract agent for delaying or slowing down the age-related cataracts in humans.

5. Grx2 interactions and protein targets

Ribonucleotide reductase has been classified as a primary target for electron donation from several members of the thioredoxin superfamily including Trx1, Trx2, Grx1, and Grx2. In DNA nucleotide synthesis, ribonucleotide reductase performs the rate-limiting step by converting ribonucleotides to deoxyribonucleotides. dNTPs are crucial for DNA reproduction and repair. Therefore, by mediating ribonucleotide reductase, Grx2 essentially allows for sustained DNA production even in times of oxidative damage. Evidence shows that Grx2 uses a monothiol mechanism to donate electrons with activity highly dependent on its cofactor, GSH, concentration.

Several studies have demonstrated Grx2's ability to protect classical mitochondrial targets, such as complex I and IV. Particularly, Wu et al. have shown that complex I and IV glutathionylation is reversed by Grx2 in retinal pigment epithelial (RPE) cells and furthermore, prevents hydrogen peroxide (H_2O_2)-induced damage [31]. UCP1 in BAT and UCP3 in skeletal muscle are involved in thermogenesis and ROS production and signaling [45, 46]. Allosteric control of these proteins relies on Grx1 and Grx2's reversible glutathionylation. *Grx2* KO mice displayed slightly significant lower body mass, accompanied by denser gonadal white adipose tissue and less reactivity to H_2O_2-induced activation of UCP3 [46]. Without Grx2, UCP3 activation ceased. TCA cycle metabolites such as succinate and pyruvate were decreased in the liver, compared to a decrease in 2-oxoglutarate in skeletal muscle [51]. 2-Oxoglutarate dehydrogenase (Ogdh) is another enzyme that has been shown to be directly regulated by ROS-induced glutathionylation. Grx2 works to deglutathionylate Ogdh, and Ogdh in return also controls mitochondrial redox balance by preserving normal ROS levels. With H_2O_2 treatment, GSH initially feeds the amplifying cascade of ROS production by Ogdh, but ROS levels are strictly controlled by Grx2's activation and direct peroxidase activities [51]. The inhibition of Grx2 leads to the limitation of Ogdh activity and could explain lower amounts of 2-oxoglutarate in *Grx2* KO mice.

Grx2 has also been implicated to interact with TR to reduce certain substrates. With the help of TR, Grx2's primary protein targets are coenzyme A (CoA) and other small molecule disulfides [7, 22, 23]. Because of Grx2's high affinity to oxidized targets, CoA mixed disulfide and glutathione disulfide were easily reduced due to TR's ability to reduce both disulfide bridges and mixed glutathione disulfide at Grx2's catalytic Cys sites [22, 52, 53]. Interestingly, there is also evidence that supports that Grx2 may also use TR as an electron donor to reduce the GSSG mitochondrial pool especially in times of GSH deficiency or mitochondrial dysfunction [26, 35].

Oxidative stress may lead to inactivation of other fundamental antioxidant enzymes. Compared to other antioxidant enzymes, Grx2 is less prone to inactivation and has higher affinity to glutathionylated substrates, making it ideal antioxidant enzyme in terms of scavenging free radicals in highly oxidizing environments [22]. Other antioxidant enzymes like TR has a sensitive selenocysteine at the active site that makes it susceptible to electrophilic oxidation and inactivation with abundant oxidative stress [54]. A study by Zhang et al. used TR inhibitors such as auranofin and 4-hydroxynonenal (HNE) to determine the effect on Grx2 activity. Seemingly, Grx2 and cytosolic Trx1 and mitochondrial Trx2 act as back-up systems to each other [23]. When Trx1 and Trx2 are compromised, Grx2 expression is increased, and vice-versa. GSH was also shown to directly reduce Trx1 and Trx2, and the addition of Grx2 increased Trx's repair and enzyme activity [7, 55]. When Grx2 overexpression was induced in HeLa cells, HeLa cells were resistant to auranofin and HNE's cytotoxic effects by promoting anti-apoptotic effects and dually reducing and activating oxidized Trx [55]. Interestingly, considering Trx2 and Grx2's principal mitochondrial localization, the two enzymes may also share common targets. Peroxiredoxin 3 (Prx3) is especially located in the mitochondria and has direct peroxidase and signaling activities. Prx3 has two catalytic cysteines, as do most of the typical members of the peroxiredoxin family, but can also be regulated by oxidation of certain structural Cys residues [56]. Hanschmann et al. were able to determine that both Grx2 and Trx2 work as electron donors for Prx3 [57]. Inhibition of either Grx2 or Trx2 does not affect the expression or activation of Prx3, reinforcing the idea that Grx2 and Trx2 may work as back-up systems for each other [7, 23, 56]. However, among dual inhibition, oxidized Prx3 accumulates in the mitochondria. Grx2 also may have functions in preventing the oxidation of ascorbate or vitamin C. Grx2 acts as an electron donor to dehydroascorbate (DHA) reductase, an enzyme responsible for the conversion of DHA into ascorbate [11]. Vitamin C can also then potentially work as an antioxidant and scavenge free radicals in the cell.

With Grx2's mitochondrial and nuclear localization, Grx2 has the potential to regulate the activity and expression of several different proteins. Unfortunately, there is limited research on Grx2 protein targets, yet screening-based assays help identify certain possibilities that may explain why Grx2 is anti-apoptotic or capable of enhancing cytoprotection in oxidative stress environments. Schutte et al. identified that cytoskeleton, chaperone, proteolysis, metabolic, and transcription factor proteins were the most likely candidates to be regulated by Grx2 [24]. Considering Grx2's effects on proliferation and migration, proteins like actin, tubulin, and dihydropyrimidinase-related protein 2 (DPYL2) with key cysteine residues are regulated by Grx2c. Chaperone proteins have multiple functions including proper folding of proteins and response to ER and oxidative stress, so heat shock proteins may be critically controlled by Grx2 enzymatic activity. Grx2's mitochondrial localization makes it a prime candidate for ensuring metabolic activity. Previous studies have highlighted Grx2's management of TCA cycle metabolites and ATP production in the cell [51, 58]. This may explain why such as glyceraldehyde-3-phosphate (G3PD), enoyl CoA hydratase (ECHM), and glycine N-methyltransferase (GNMT) are specific Grx2 targets [24]. Moreover, ubiquitination is closely linked with oxidative stress and autophagy, which may explain why 26S proteasome subunits and certain E3 ubiquitin ligases may be possible Grx2 targets [24]. Regarding Grx2's nuclear localization especially in HeLa cells, this definitely suggests that Grx2 may be vitally important in activating crucial transcription factors or signaling pathways [2, 11, 50]. The screening-based assay in Schutte's study highlighted multiple key transcription factors including ruvB-like 1 (RUVB1), elongation factor tu (EFTU), and heterogeneous nuclear ribonucleoprotein F (HNRPF) [24]. Although some of these targets have not been confirmed, it still provides insightful information about how Grx2 may facilitate cellular survival and protection.

6. Conclusion: Grx2's anti-apoptotic and cytoprotective roles may make it a potential drug target to treat oxidative stress-induced diseases

Despite being overlooked, research throughout the years has portrayed Grx2 as a capable and robust antioxidant enzyme through direct detoxification of free radicals, cytoprotection, and anti-apoptotic abilities [5, 23]. Grx2's primary mitochondrial and nuclear localization highlights its important role in balancing pro- and anti-apoptotic signals, maintaining metabolic activity, and preventing cellular dysfunction. Furthermore, Grx2 remains inactive as a dimer until oxidative stress is present, indicating that Grx2 expression can be finely manipulated [17, 21]. Several degenerative and oxidative stress-induced diseases involve exacerbated cell death, and Grx2's potent anti-apoptotic effects make it a marketable drug target to combat oxidative stress and injury [2, 5, 26]. More research needs to be focused on novel drug discovery research to find inhibitors for the treatment and drug sensitization of cancer and activators for the treatment of oxidative stress and degenerative diseases.

Acknowledgements

This research is supported in part by Bright Focus Foundation for Macular Degeneration (Grant No. M2015180) and California Table Grape Commission (Grant to Hongli Wu).

Author details

Christy Xavier[1], Xiaobin Liu[1], Yang Liu[2] and Hongli Wu[1,2*]

*Address all correspondence to: hongli.wu@unthsc.edu

1 Pharmaceutical Sciences, College of Pharmacy, University of North Texas Health Science Center, University of North Texas System, Fort Worth, Texas, United States

2 North Texas Eye Research Institute, University of North Texas Health Science Center, Fort Worth, Texas, United States

References

[1] Murphy MP. How mitochondria produce reactive oxygen species. Biochemical Journal. 2009;**417**(Pt 1):1-13

[2] Hybertson BM, Gao B, Bose SK, McCord JM. Oxidative stress in health and disease: The therapeutic potential of Nrf2 activation. Molecular Aspects of Medicine. 2011; **32**(4-6):234-246

[3] Dalle-Donne I, Rossi R, Giustarini D, Colombo R, Milzani A. S-glutathionylation in protein redox regulation. Free Radical Biology and Medicine. 2007;**43**(6):883-898

[4] Mailloux RJ, Willmore WG. S-glutathionylation reactions in mitochondrial function and disease. Frontiers in Cell and Developmental Biology. 2014;**2**:68

[5] Ghezzi P. Protein glutathionylation in health and disease. Biochimica et Biophysica Acta. 2013;**1830**(5):3165-3172

[6] Holmgren A, Johansson C, Berndt C, Lonn ME, Hudemann C, Lillig CH. Thiol redox control via thioredoxin and glutaredoxin systems. Biochemical Society Transactions. 2005;**33**(Pt 6):1375-1377

[7] Kalinina EV, Chernov NN, Saprin AN. Involvement of thio-, peroxi-, and glutaredoxins in cellular redox-dependent processes. Biochemistry. Biokhimiia. 2008;**73**(13):1493-1510

[8] Mates JM, Perez-Gomez C, Nunez de Castro I. Antioxidant enzymes and human diseases. Clinical Biochemistry. 1999;**32**(8):595-603

[9] Holmgren A. Glutathione-dependent synthesis of deoxyribonucleotides. Purification and characterization of glutaredoxin from *Escherichia coli*. The Journal of Biological Chemistry. 1979;**254**(9):3664-3671

[10] Lonn ME, Hudemann C, Berndt C, Cherkasov V, Capani F, Holmgren A, Lillig CH. Expression pattern of human glutaredoxin 2 isoforms: Identification and characterization of two testis/cancer cell-specific isoforms. Antioxidants & Redox Signaling. 2008; **10**(3):547-557

[11] Lundberg M, Johansson C, Chandra J, Enoksson M, Jacobsson G, Ljung J, Johansson M, Holmgren A. Cloning and expression of a novel human glutaredoxin (Grx2) with mitochondrial and nuclear isoforms. Journal of Biological Chemistry. 2001;**276**(28):26269-26275

[12] Hudemann C, Lonn ME, Godoy JR, Zahedi Avval F, Capani F, Holmgren A, Lillig CH. Identification, expression pattern, and characterization of mouse glutaredoxin 2 isoforms. Antioxidants & Redox Signaling. 2009;**11**(1):1-14

[13] Jurado J, Prieto-Álamo M-J, Madrid-Rísquez J, Pueyo C. Absolute gene expression patterns of thioredoxin and glutaredoxin redox systems in mouse. Journal of Biological Chemistry. 2003;**278**(46):45546-45554

[14] Mollbrink A, Jawad R, Vlamis-Gardikas A, Edenvik P, Isaksson B, Danielsson O, Stal P, Fernandes AP. Expression of thioredoxins and glutaredoxins in human hepatocellular carcinoma: Correlation to cell proliferation, tumor size and metabolic syndrome. International Journal of Immunopathology and Pharmacology. 2014;**27**(2):169-183

[15] Fernandes AP, Capitanio A, Selenius M, Brodin O, Rundlof AK, Bjornstedt M. Expression profiles of thioredoxin family proteins in human lung cancer tissue: Correlation with proliferation and differentiation. Histopathology. 2009;**55**(3):313-320

[16] Godoy JR, Oesteritz S, Hanschmann EM, Ockenga W, Ackermann W, Lillig CH. Segment-specific overexpression of redoxins after renal ischemia and reperfusion: Protective roles of glutaredoxin 2, peroxiredoxin 3, and peroxiredoxin 6. Free Radical Biology & Medicine. 2011;**51**(2):552-561

[17] Hashemy SI, Johansson C, Berndt C, Lillig CH, Holmgren A. Oxidation and S-nitro-sylation of cysteines in human cytosolic and mitochondrial glutaredoxins: Effects on structure and activity. The Journal of Biological Chemistry. 2007;**282**(19):14428-14436

[18] Upadhyaya B, Tian X, Wu H, Lou MF. Expression and distribution of thiol-regulating enzyme glutaredoxin 2 (GRX2) in porcine ocular tissues. Experimental Eye Research. 2015;**130**:58-65

[19] Berndt C, Hudemann C, Hanschmann EM, Axelsson R, Holmgren A, Lillig CH. How does iron-sulfur cluster coordination regulate the activity of human glutaredoxin 2? Antioxidants & Redox Signaling. 2007;**9**(1):151-157

[20] Lillig CH, Berndt C, Vergnolle O, Lonn ME, Hudemann C, Bill E, Holmgren A. Charac-terization of human glutaredoxin 2 as iron-sulfur protein: A possible role as redox sen-sor. Proceedings of the National Academy of Sciences of the United States of America. 2005;**102**(23):8168-8173

[21] Mitra S, Elliott SJ. Oxidative disassembly of the [2Fe-2S] cluster of human Grx2 and redox regulation in the mitochondria. Biochemistry. 2009;**48**(18):3813-3815

[22] Johansson C, Lillig CH, Holmgren A. Human mitochondrial glutaredoxin reduces S-glutathionylated proteins with high affinity accepting electrons from either glutathi-one or thioredoxin reductase. The Journal of Biological Chemistry. 2004;**279**(9):7537-7543

[23] Fernandes AP, Holmgren A. Glutaredoxins: Glutathione-dependent redox enzymes with functions far beyond a simple thioredoxin backup system. Antioxidants & Redox Signaling. 2004;**6**(1):63-74

[24] Schutte LD, Baumeister S, Weis B, Hudemann C, Hanschmann EM, Lillig CH. Iden-tification of potential protein dithiol-disulfide substrates of mammalian Grx2. Biochi-mica et Biophysica Acta. 2013;**1830**(11):4999-5005

[25] Sagemark J, Elgan TH, Burglin TR, Johansson C, Holmgren A, Berndt KD. Redox prop-erties and evolution of human glutaredoxins. Proteins. 2007;**68**(4):879-892

[26] Higgins GC, Beart PM, Shin YS, Chen MJ, Cheung NS, Nagley P. Oxidative stress: Emerging mitochondrial and cellular themes and variations in neuronal injury. Journal of Alzheimer's disease : JAD. 2010;**20**(Suppl 2):S453-S473

[27] Ferri A, Fiorenzo P, Nencini M, Cozzolino M, Pesaresi MG, Valle C, Sepe S, Moreno S, Carri MT. Glutaredoxin 2 prevents aggregation of mutant SOD1 in mitochondria and abolishes its toxicity. Human Molecular Genetics. 2010;**19**(22):4529-4542

[28] Mannarino SC, Vilela LF, Brasil AA, Aranha JN, Moradas-Ferreira P, Pereira MD, Costa V, Eleutherio EC. Requirement of glutathione for Sod1 activation during lifespan exten-sion. Yeast (Chichester, England). 2011;**28**(1):19-25

[29] Arodin L, Miranda-Vizuete A, Swoboda P, Fernandes AP. Protective effects of the thio-redoxin and glutaredoxin systems in dopamine-induced cell death. Free Radical Biology and Medicine. 2014;**73**:328-336

[30] Karunakaran S, Saeed U, Ramakrishnan S, Koumar RC, Ravindranath V. Constitutive expression and functional characterization of mitochondrial glutaredoxin (Grx2) in mouse and human brain. Brain Research. 2007;**1185**:8-17

[31] Wu H, Xing K, Lou MF. Glutaredoxin 2 prevents H_2O_2-induced cell apoptosis by pro-tecting complex I activity in the mitochondria. Biochimica et Biophysica Acta. 2010;**1797**(10):1705-1715

[32] Romero JI, Hanschmann EM, Gellert M, Eitner S, Holubiec MI, Blanco-Calvo E, Lillig CH, Capani F. Thioredoxin 1 and glutaredoxin 2 contribute to maintain the phenotype and integrity of neurons following perinatal asphyxia. Biochimica et Biophysica Acta. 2015;**1850**(6):1274-1285

[33] Bräutigam L, Schütte LD, Godoy JR, Prozorovski T, Gellert M, Hauptmann G, Holmgren A, Lillig CH, Berndt C. Vertebrate-specific glutaredoxin is essential for brain develop-ment. Proceedings of the National Academy of Sciences. 2011;**108**(51):20532-20537

[34] Gellert M, Venz S, Mitlöhner J, Cott C, Hanschmann E-M, Lillig CH. Identification of a dithiol-disulfide switch in collapsin response mediator protein 2 (CRMP2) that is toggled in a model of neuronal differentiation. Journal of Biological Chemistry. 2013;**288**(49):35117-35125

[35] Fernando MR, Lechner JM, Lofgren S, Gladyshev VN, Lou MF. Mitochondrial thioltrans-ferase (glutaredoxin 2) has GSH-dependent and thioredoxin reductase-dependent perox-idase activities in vitro and in lens epithelial cells. FASEB Journal. 2006;**20**(14):2645-2647

[36] Wu H, Lin L, Giblin F, Ho YS, Lou MF. Glutaredoxin 2 knockout increases sensitiv-ity to oxidative stress in mouse lens epithelial cells. Free Radical Biology & Medicine. 2011;**51**(11):2108-2117

[37] Wu H, Yu Y, David L, Ho Y-S, Lou MF. Glutaredoxin 2 (Grx2) gene deletion induces early onset of age-dependent cataracts in mice. Journal of Biological Chemistry. 2014;**289**(52):36125-36139

[38] Sordahl LA. Role of mitochondria in heart cell function. Texas Reports on Biology and Medicine. 1979;**39**:5-18

[39] Berndt C, Poschmann G, Stühler K, Holmgren A, Bräutigam L. Zebrafish heart develop-ment is regulated via glutaredoxin 2 dependent migration and survival of neural crest cells. Redox Biology. 2014;**2**:673-678

[40] Pai HV, Starke DW, Lesnefsky EJ, Hoppel CL, Mieyal JJ. What is the functional sig-nificance of the unique location of glutaredoxin 1 (GRx1) in the intermembrane space of mitochondria? Antioxidants & Redox Signaling. 2007;**9**(11):2027-2033

[41] Bräutigam L, Jensen LDE, Poschmann G, Nyström S, Bannenberg S, Dreij K, Lepka K, Prozorovski T, Montano SJ, Aktas O, Uhlén P, Stühler K, Cao Y, Holmgren A, Berndt C.

Glutaredoxin regulates vascular development by reversible glutathionylation of sirtuin 1. Proceedings of the National Academy of Sciences. 2013;**110**(50):20057-20062

[42] Mailloux RJ, Xuan JY, McBride S, Maharsy W, Thorn S, Holterman CE, Kennedy CRJ, Rippstein P, deKemp R, da Silva J, Nemer M, Lou M, Harper M-E. Glutaredoxin-2 Is required to control oxidative phosphorylation in cardiac muscle by mediating deglutathionylation reactions. Journal of Biological Chemistry. 2014;**289**(21):14812-14828

[43] Gao X-H, Qanungo S, Pai HV, Starke DW, Steller KM, Fujioka H, Lesnefsky EJ, Kerner J, Rosca MG, Hoppel CL, Mieyal JJ. Aging-dependent changes in rat heart mitochondrial glutaredoxins—Implications for redox regulation. Redox Biology. 2013;**1**(1):586-598

[44] Sakai J, Li J, Subramanian KK, Mondal S, Bajrami B, Hattori H, Jia Y, Dickinson BC, Zhong J, Ye K, Chang CJ, Ho Y-S, Zhou J, Luo HR. Reactive oxygen species (ROS)-induced actin glutathionylation controls actin dynamics in neutrophils. Immunity. 2012;**37**(6):1037-1049

[45] Mailloux RJ, Adjeitey CN, Xuan JY, Harper ME. Crucial yet divergent roles of mitochondrial redox state in skeletal muscle vs. brown adipose tissue energetics. FASEB Journal. 2012;**26**(1):363-375

[46] Mailloux RJ, Xuan JY, Beauchamp B, Jui L, Lou M, Harper ME. Glutaredoxin-2 is required to control proton leak through uncoupling protein-3. The Journal of Biological Chemistry. 2013;**288**(12):8365-8379

[47] Stuart JA, Cadenas S, Jekabsons MB, Roussel D, Brand MD. Mitochondrial proton leak and the uncoupling protein 1 homologues. Biochimica et Biophysica Acta (BBA): Bioenergetics. 2001;**1504**(1):144-158

[48] Enoksson M, Fernandes AP, Prast S, Lillig CH, Holmgren A, Orrenius S. Overexpression of glutaredoxin 2 attenuates apoptosis by preventing cytochrome c release. Biochemical and Biophysical Research Communications. 2005;**327**(3):774-779

[49] Elmore S. Apoptosis: A review of programmed cell death. Toxicologic Pathology. 2007;**35**(4):495-516

[50] Lillig CH, Lonn ME, Enoksson M, Fernandes AP, Holmgren A. Short interfering RNA-mediated silencing of glutaredoxin 2 increases the sensitivity of HeLa cells toward doxorubicin and phenylarsine oxide. Proceedings of the National Academy of Sciences of the United States of America. 2004;**101**(36):13227-13232

[51] Mailloux RJ, Craig Ayre D, Christian SL. Induction of mitochondrial reactive oxygen species production by GSH mediated S-glutathionylation of 2-oxoglutarate dehydrogenase. Redox Biology. 2016;**8**:285-297

[52] Lillig CH, Prior A, Schwenn JD, Aslund F, Ritz D, Vlamis-Gardikas A, Holmgren A. New thioredoxins and glutaredoxins as electron donors of 3'-phosphoadenylylsulfate reductase. The Journal of Biological Chemistry. 1999;**274**(12):7695-7698

[53] Zahedi Avval F, Holmgren A. Molecular mechanisms of thioredoxin and glutaredoxin as hydrogen donors for mammalian S phase ribonucleotide reductase. The Journal of Biological Chemistry. 2009;**284**(13):8233-8240

[54] Nauser T, Steinmann D, Grassi G, Koppenol WH. Why selenocysteine replaces cysteine in thioredoxin reductase: A radical hypothesis. Biochemistry. 2014;**53**(30):5017-5022

[55] Zhang H, Du Y, Zhang X, Lu J, Holmgren A. Glutaredoxin 2 reduces both thioredoxin 2 and thioredoxin 1 and protects cells from apoptosis induced by auranofin and 4-hydroxynonenal. Antioxidants & Redox Signaling. 2014;**21**(5):669-681

[56] Chang T-S, Cho C-S, Park S, Yu S, Kang SW, Rhee SG. Peroxiredoxin III, a mitochondrion-specific peroxidase, regulates apoptotic signaling by mitochondria. The Journal of Biological Chemistry. 2004;**279**(40):41975-41984

[57] Hanschmann EM, Lonn ME, Schutte LD, Funke M, Godoy JR, Eitner S, Hudemann C, Lillig CH. Both thioredoxin 2 and glutaredoxin 2 contribute to the reduction of the mitochondrial 2-Cys peroxiredoxin Prx3. The Journal of Biological Chemistry. 2010;**285**(52):40699-40705

[58] Sagemark J, Elgán TH, Bürglin TR, Johansson C, Holmgren A, Berndt KD. Redox properties and evolution of human glutaredoxins. Proteins: Structure, Function, and Bioinformatics. 2007;**68**(4):879-892

Glutathione and Oxidative Stress

Impact of Oxidative Changes and Possible Effects of Genetics Polymorphisms of Glutathione S-Transferase in Diabetics Patients with Complications

Laura Raniere Borges dos Anjos,
Ana Cristina Silva Rebelo,
Gustavo Rodrigues Pedrino,
Rodrigo da Silva Santos and
Angela Adamski da Silva Reis

Additional information is available at the end of the chapter

http://dx.doi.org/10.5772/intechopen.76222

Abstract

Pancreatic β cells are more sensitive to cytotoxic stress than several other cells due to the expression of very low levels of antioxidant enzymes. Glutathione-S-transferase (GST) is a detoxification enzyme essential for a cellular protection against oxidative damage. Thus, the objective of this chapter is to verify the impact of the hypothesis of all effects of Glutathione S-transferase polymorphism in patients with diabetic complications. Diabetic nephropathy (DN) is the main secondary complication of diabetes mellitus (DM). Notably, the expression of GST genes has been described in different variations as ethnic populations. Some studies have suggested association between genetic polymorphism for GSTM1, GSTT1 and *GSTP1* and DN, but others do not. The results are still inconsistent and, therefore, more studies are needed to be performed.

Keywords: GST, diabetic nephropathy, diabetes, polymorphism, glutathione S-transferase

1. Introduction

Diabetes mellitus (DM) is defined as a heterogeneous group of metabolic disorders characterized by an unusual hyperglycemia resulting from defects in insulin action and/or secretion. An epidemic of DM is underway as result of population growth and aging, increased urbanization,

IntechOpen

prevalence of obesity and sedentary lifestyle [1]. It is estimated that currently about 415 million individuals are diagnosed with DM worldwide and it is predicted that by 2040 these records will reach the range of 672 million [2].

Although the survival of these patients has increased in recent decades, it is known that the morbidity resulting from complications affecting the small blood vessels (microvascular) or large (macrovascular) arteries is very significant. These complications may occur as consequence of hyperglycemia that favors inadequate cellular metabolism and excessive production of reactive oxygen species (ROS). The etiopathogenesis of DM is not fully elucidated, but it is suggested that genetic and environmental factors are involved in this metabolic disorder and, in this sense, oxidative stress becomes one of the important pathways for this understanding [3].

Human cells produce many antioxidants in attempt to protect cells against damage caused by toxins from the environment. The main endogenous antioxidant defense is provided by glutathione (GSH). GSH is a linear tripeptide that arouses scientific interest because it performs multiple functions via glutathione S-transferase (GST). GSTs are one of the most important groups of phase II enzymes. It is reported that these enzymes are induced, as a protective mechanism (detoxification), under conditions of oxidative stress. GST polymorphisms were associated with increased or decreased susceptibility of various diseases, such as cancer, cardiac diseases, about everything diabetes and yours complications [4].

Some important members of the GST family stand out for having different polymorphisms between these GST mu 1 (GST M1) and GST theta 1 (GST T1) and GST Pi 1 (GSTP 1). It is reported that these GSTs subtypes are involved in the development of DM and its complications [5], so it is important to understand the impact of these oxidative changes and the possible effects of genetic polymorphisms of GSTs in diabetic patients [6].

2. Oxidative stress

Living aerobic organisms have an intracellular environment in which important biological molecules are in equilibrium, and oxidative metabolism and redox homeostasis are in sync. In these organisms, oxidative phosphorylation is a vital step of metabolism [7].

This metabolic pathway uses the energy generated by NAD^+ oxy-reduction reactions in NADH and produces adenosine triphosphate (ATP) molecules capable of storing energy for immediate consumption [8]. As consequence, free radicals are produced naturally and continuously [9]. It is important to note that the mechanism of free radical generation can also occur in cell membranes and cytoplasm with the participation of transition metals such as iron and copper [10].

In the body, free radicals can act in a beneficial way during the immune response, destroying invading pathogens and modulating the excessive inflammatory response, however, their excess may cause deleterious effects to the organism. Normally, in healthy living organisms, there is a balance between the production of free radicals and antioxidant systems [11].

The imbalance between the production and the antioxidant defense capacity of the organism is called oxidative stress [12]. The cellular effects of this hostile environment depend on factors such as cell type, presence of surface receptors, mechanism of transduction and levels of antioxidants [7]. But it is known that prolonged exposure to oxidative stress can damage cellular components (proteins, lipids and DNA) [13], contribute to cellular aging [14], and play an important role in the pathogenesis of cancer, atherosclerosis, Parkinson, Alzheimer's and various chronic diseases such as diabetes mellitus and its complications [15–20].

3. The biological role of glutathione and glutathione S-transferases in oxidative stress

Numerous studies have shown that in order to avoid prolonged exposure to ROS produced during oxidative stress, the body has a very efficient antioxidant defense system. Glutathione S-transferases (GSTs) and glutathione (GSH) enzymes are part of this line of defense [21].

Glutathione (GSH) is a low molecular weight thiol found in all tissues, primarily in aerobic organisms. Also known as L-gamma-glutamyl-L-cysteinyl-glycine, GSH is a linear tripeptide consisting of three amino acids: glutamic acid, cysteine and glycine (**Figure 1**). Between the γ-glutamyl moiety and the free α-carboxylate group, there is a γ-peptide bond which, although unusual, prevents the hydrolysis of GSH by cellular peptidases [22].

In homeostasis conditions, GSH is the most efficient physiological reducing agent with the highest bioavailability (~ 10 mM) in the intracellular environment where it is synthesized,

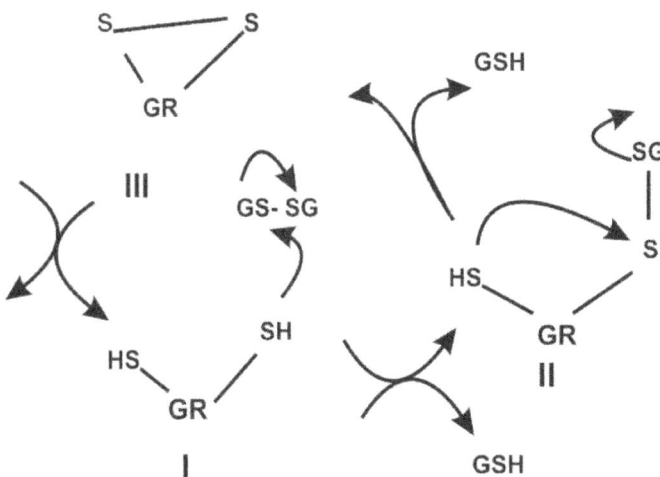

Figure 1. Schematic representation of the GSSG reduction cycle by GR.

except in epithelial cells [23]. Its synthesis occurs in two phases and counts on the action of two enzymes: γ-glutamyl-cysteine-synthetase and glutathione-synthetase [24].

In the first phase, the γ-glutamyl-cysteine-synthetase enzyme favors the formation of the peptide bond between glutamic acid and cysteine, thus forming the dipeptide γ-L-glutamyl-L-cysteine [25]. In the second phase, the enzyme glutathione synthetase binds the newly formed dipeptide to glycine, giving rise to GSH which is distributed through the bloodstream and then brought to the tissues. In both phases, consumption of ATP and Mg^{+2} occurs. The regulation of the enzyme γ-glutamyl-cysteine-synthetase is done, by negative feedback, when the GSH itself begins to be formed. This regulatory mechanism ensures that, in normal conditions, the excessive production of GSH or the intermediate γ-L-glutamyl-L-cysteine does not occur (**Figure 2**) [22, 24].

An alternative route is activated in situations where conversion of γ-glutamyl-L-cysteine into GSH is insufficient. In this case, the enzyme γ-glutamylcyclotransferase catalyzes the conversion of γ-glutamyl-L-cysteine to 5-oxoproline, favoring the occurrence of 5-oxoprolinuria, chronic metabolic acidosis and neurological disorders (**Figure 2**) [22].

During the reaction catalyzed by γ-glutamylcysteine synthetase, activation of butionin sulfoximine (BSO), an inhibitor of GSH biosynthesis, may occur. Studies suggest that this suppression of GSH by BSO may be a rather efficient strategy in cancer therapy since, during this process, there is an increase in the sensitivity of cells to ionizing radiation and to cytostatic drugs, making them more susceptible to treatment. However, the disadvantage of this technique is that the toxic effect to normal cells has potency detrimental to the individual. An alternative to limit this toxicity would be the use of localized irradiation or the topical application of cytostatic drugs, but other studies are being carried out [26].

Glutathione can be found in the intracellular medium in its reduced (GSH) or oxidized form (GSSG, dimerized form of GSH) and the GSH/GSSG ratio determines the redox state of biological systems. This is because glutathione performs a cytotoxic and genotoxic inactivation of xenobiotics and consequently promotes detoxification and cellular protection against oxidative stress and additional damage [27].

The cellular detoxification process is divided into three distinct but related phases. In phases I and II, the xenobiotic is transformed into a more soluble and less toxic product and, in phase III, are transported for cellular excretion. It is noteworthy that the efficiency of phase II depends on the action of enzymes called glutathione S-transferases (GSTs) [22].

The GSTs belong to a superfamily of multigenic enzymes that catalyze the nucleophilic attack of the reduced form of Glutathione (GSH) to compounds that present a carbon, a nitrogen or an electrophilic sulfur atom [21]. Under natural conditions, GSTs are generally found in the biological environment as homo or heterodimers. Each dimer contains two active sites with independent activities. Each site has at least two binding regions: one specific for glutathione (GSH) and the other, with less specificity, for the electrophiles (alkyl halides, epoxides, quinones, iminoquinones, aldehydes, ketones, lactones and esters, halides of aryl and aromatic nitro) [22, 28].

Mammalian GSTs are divided into families according to their location: cytosolic, mitochondrial and microsomal. The cytosolic and mitochondrial GST enzymes are soluble, unlike the microsomal GSTs that are associated with the membrane [29]. This latter family is generally

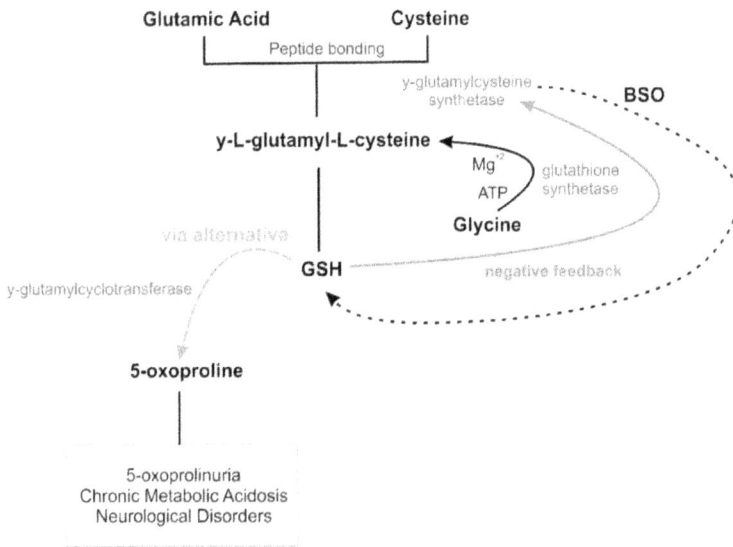

Figure 2. Scheme representing the biosynthesis and mechanism of regulation of glutathione (GSH). BSO, butionin sulfoximine; Mg^{+2}, magnesium; ATP, adenosine triphosphate.

involved in the metabolism of eicosanoids and glutathione (GSH), thus being referred to as MAPEG (membrane-associated proteins in eicosanoid and glutathione metabolism) [22]. It is important to note that other families of GSTs, absent in mammals, are also described in the literature. Cytosolic and mitochondrial GSTs are expressed in different isoforms and therefore divided into classes according to the amino acid and/or nucleotide sequence, immunological properties, enzymatic kinetic parameters and/or tertiary and quaternary structure [22, 29, 30].

Based on the similarity of the amino acid sequence, GSTS found in the cytosol are called α (GSTA), μ (GSTM), θ (GSTT), π (GSTP), σ (GSTS), omega (GSTO), and zeta (GSTZ) [28, 31]. The mitochondrial GST is known as kappa (GSTK) [31]. Mammalian cytosolic GSTs are all dimeric and contain 199–244 amino acid residues in their primary structures. Mitochondrial GSTs are also dimeric proteins and their subunits typically have 226 amino acid residues. Each of these enzymes differs in their functionality [22, 33]. It is suggested that in humans, the expression of these enzymes is uniform and independent of the cell type. GSTs have long been described as originating from mitochondria; however, recent studies indicate the presence of mitochondrial GSTs in peroxisomes. These findings reinforce their participation in the detoxification processes of by-products of β-oxidation of fatty acids [22].

During the detoxification process, the GSTs catalyze the conjugation of xenobiotics with endogenous substrates, mainly GSH. This conjugate is recognized by specific transporters and is carried to the intercellular medium where it undergoes action of γ-glutamyl transpeptidase (γGT) which removes the glutamic acid residue [32]. In sequencing, the dipeptidases remove the glycine residue, leaving only the cysteine residue associated with the xenobiotic. The

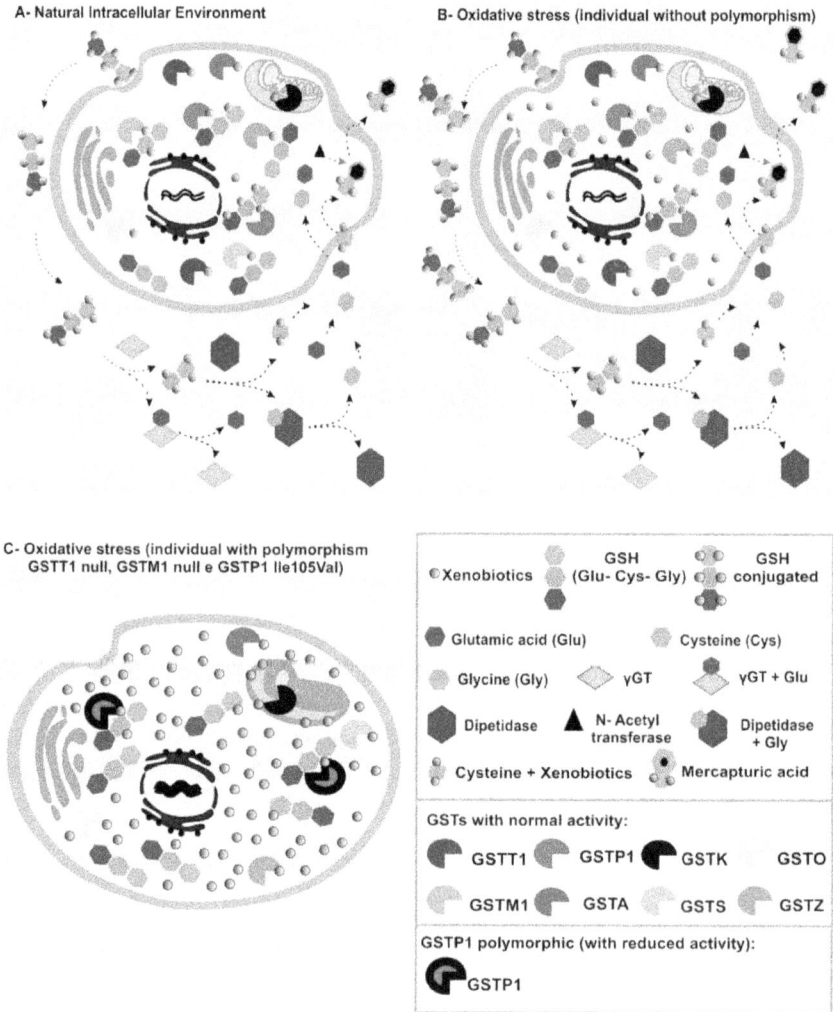

Figure 3. Schematic representation of main glutathiones S-Transferases correlated with oxidative stress in different biological conditions. A, normal intracellular environment; B, oxidative stress in an individual without polymorphism; C, oxidative stress in an individual with oxidative stress.

amino group of the cysteine residue is then acetylated by the intracellular N-acetyltransferase enzymes and thereby forms the mercapturic acid which, depending on its characteristics, is rapidly led to circulation, bile, urine or metabolized until it is eliminated (**Figure 3**) [22].

Once free, glutamate and glycine are reabsorbed by the cell and used in the regeneration of GSH through the catalytic cycle. In this stage of regeneration, three groups of enzymes are important: glutathione oxidase (GO) and glutathione peroxidase (GSH-Px), which catalyze the oxidation of GSH to GSSG, and the enzyme glutathione reductase (GR) that is responsible for the regeneration of GSH, from GSSG, in the presence of NADPH [33].

It is important to highlight that this mechanism of detoxification via glutathione represents a fundamental biochemical evolution for the survival and guarantee of the perpetuation of many species and, although a co-transport mechanism without conjugate envelopment with glutathione has been proposed, there is no evidence experimental models that validate this model [22].

4. Oxidative alterations and the pathophysiology of diabetes and its complications

Many studies suggest that patients with diabetes present alterations in the levels of reactive oxygen species (ROS), a type of free radical whose electron is centered in the oxygen atoms [34]. This fact is justified by the toxic character of the persistent excess of glucose in the organism that ends up promoting glycation of proteins, hyperosmolarity and increase in the levels of sorbitol inside the cells [35].

Glucose is a vital source of energy for cells, and their serum levels are controlled by various organs such as intestine, liver, pancreas, skeletal muscle, adipose tissue and kidneys [36]. This regulation is facilitated by the action of hormones (glucagon and insulin), central and peripheral nervous system, as well as metabolic requirements of the body [37].

DM is defined as a heterogeneous group of metabolic disorders characterized by unusual hyperglycemia resulting from defects in insulin production and/or action [1]. In this situation, to revert the toxicity of excess glucose, this component undergoes auto-oxidation and, as consequence, ROS are generated (**Figure 4**) [37].

During auto-oxidation, excess glucose binds (protein glycation) [37] to lysine and valine residues in tissue and plasma proteins. This interaction results in the formation of Schiff's base, a labile or unstable compound that spontaneously transforms into ketoamine (glycated hemoglobin) through the *Amadori* rearrangement [35].

These oxidation and rearrangement processes, followed by further dehydration and fragmentation of *Amadori* product, promote the formation of advanced glycation end products (AGEs) (**Figure 4**) and generate other compounds with chemically active carbonyl groups. These compounds favor the oxidative stress that affects β cells of the pancreas, responsible for synthesizing and secreting insulin [8, 38].

Accumulated AGEs bind to membrane receptors on endothelial cells and promote the onset of tissue damage and the activation of the proinflammatory pathway that involves the NFκB transcription factor responsible for regulating the expression of other inflammatory cytokines (**Figure 4**) [37].

Moreover, the chronicity of this hostile environment causes the deactivation of the nitric oxide vasodilator (NO) formed by the endothelial cells [38]. This compromises the relaxation of vascular smooth muscle cells and has a degenerative effect on the vessels causing tissue death [34] and favoring the development of microvascular complications of diabetes, such as diabetic nephropathy (DN) (**Figure 4**).

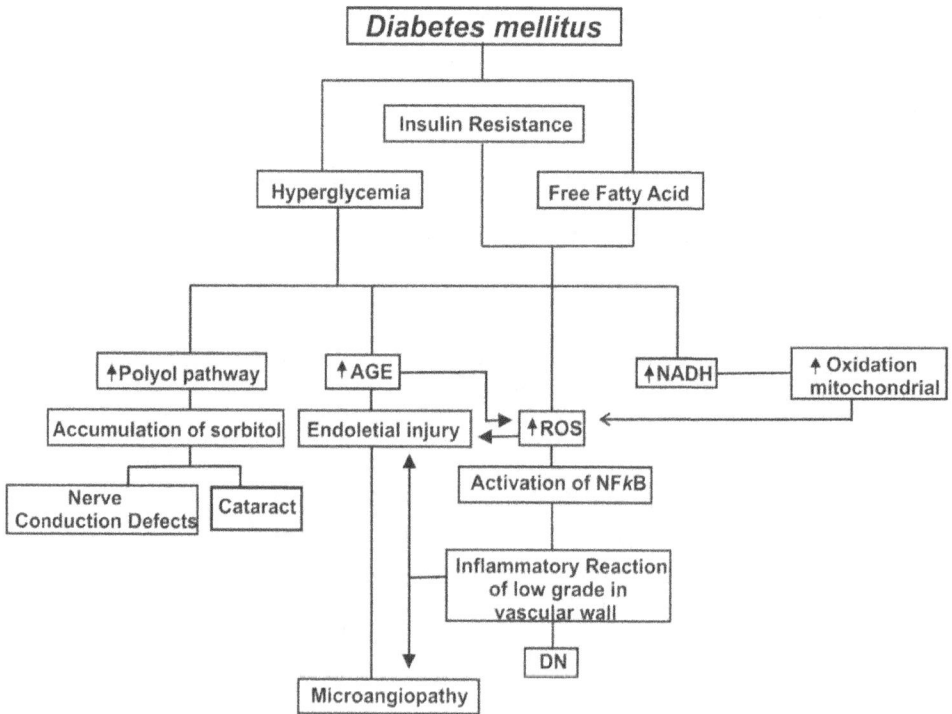

Figure 4. Main complications of *Diabetes mellitus*. NADH, nicotinamide and adenine dinucleotide; ROS, oxygen-reactive species; AGE, advanced glycation end product; DN, diabetic neuropathy.

ROS, generated by hyperglycemia, also interfere with other biochemical pathways [39]. The Krebs cycle, which, due to oxidative stress, favors the increase of the number of proton donors in the mitochondria, the main source of free radicals [37, 40]. This generates an even greater accumulation of free radicals, mainly superoxide (O_2^-) and hydroxyl compounds (OH^-) [41]. This mitochondrial production is the primary cause of long-term complications of diabetes.

The cascade signaling also suffers from oxidative stress in that it affects the activation of protein kinase C (PKC) [37], a serine/threonine kinase pathway that forms part of the mitogenic protein kinase (MAPK) [42] and plays an important role in several intracellular processes such as signal transduction, response to specific hormonal, neuronal and growth factor stimuli [28, 40].

Furthermore, hyperglycemia increases the NADH/NAD$^+$ ratio and decreases the NADPH/NADP$^+$ ratio (**Figure 4**). The substrates of this alteration are directed to the polyol pathway, which, at normal glucose concentrations, is not active [38]. In excess, in the polyol pathway, glucose is reduced to sorbitol, an osmotically active compound [37]. These disorders result in changes in redox homeostasis and in a variety of known effects for pathogenesis and progression of diabetes.

The accumulation of sorbitol in the ocular tissue, for example, contributes to the development of diabetic cataracts (**Figure 4**). In nerve tissue, high concentrations of this component decrease the uptake of myoinositol and inhibit ATPase Na$^+$/K$^+$ from the membrane, thus affecting nerve function (**Figure 4**). The accumulation of sorbitol associated with reduced hypoxia and blood flow in the nervous tissue favors the development of diabetic neuropathy [37]. This hyperglycemia may also alter gene and protein expression, endothelial cell permeability, and depletion of antioxidant molecules, including Glutathione S-transferases (GSTs), which play an important role in the cellular detoxification process [37, 41, 43–45].

5. Impact of genetic polymorphism on GSTs for patients with microvascular diabetic complications

Diabetic nephropathy (DN) is the main secondary complication of diabetes. Associated with an increased risk for cardiovascular disease and high mortality rates, DN is the leading kidney disease worldwide. Approximately 40% of diabetic patients are affected by this microvascular complication [46].

The mechanisms related to the development of DN are unclear and probably involve a number of dynamic events occurring early and with the progression of diabetes. It is known that the clinical characterization of this pathology is preceded by an established morphological renal lesion that results in imbalance of normal renal homeostasis [47]. These lesions are triggered by functional and metabolic changes. A common metabolic manifestation in the body of a diabetic individual is the picture of oxidative stress [31].

There are several factors that are involved in generating oxidative stress during diabetes. There is strong evidence that hyperglycemia results in the activation of PKC in diabetic glomeruli and, as a consequence, mesangial expansion, glomerular basement membrane thickening, endothelial cell dysfunction leading to diabetic renal disease, inflammation, apoptosis [48–50]. Diabetic renal disease, on the other hand, intensifies the formation and activation of ROS, worsening renal disease [51].

Considering that, in situations of oxidative stress, GSTs play an important role in cellular detoxification, studies of polymorphisms in the genes encoding these enzymes have been gaining prominence and arousing curiosity about a possible association with the susceptibility of this complication [52–54]. In this context, the deletions of GSTM1 and GSTT1 together with the GSTP1 Ile105Val polymorphism are among the most studied isoforms in the GSTs group [55, 56].

It is described that individuals with GSTM1 deletion polymorphisms are unable to produce the GSTM1 protein. On the other hand, the conversion of adenine to guanine at position 313 at codon 105 in the GSTP1 gene causes the amino acid isoleucine (Ile) to be replaced with valine (val), which results in a lower activity of this isoform [56].

In the last decade, some investigations have made DM associations and their complications with the genetic polymorphism in GSTs. Notably, the expression of the GST gene has been described in different variations among ethnic populations. Studies with Egyptian children and adolescents, for example, show that the null genotype of GSTT1 conferred a 4.2-fold increased risk for the occurrence of DM, and in this case, associations with some biochemical variables and laboratory data were also observed (lipid profile and HbA1c). In this study, no investigation was performed when susceptibility to DN; however, the results are clear and show that gene polymorphisms encoding GSTs are associated with the development of type 1 DM and disease-related risk factors [31].

More specific studies addressing end-stage renal failure developed as a complication of DM show that this secondary complication is more common in the Asian population than in the UK population. In addition, the data are consistent and indicate that all patients of Asian origin who developed end-stage renal failure had non-insulin-dependent diabetes [57].

A meta-analysis performed by *Saadat* (2017) [58] brought together 18 studies with a total of 5483 subjects (healthy and diabetic). Overall analysis did not indicate a significant association between *GSTP1* and type 2 DM polymorphisms. Subgroup analyzes stratified by ethnicity, year of publication, and sample size also did not reveal a significant association between study polymorphism and DM2 risk.

In contrast, another meta-analysis by *Orlewski* and *Orlewska* (2015) [29] reports strong evidence of association between the genes glutathione-S-transferase (GST) and diabetic nephropathy (DN) polymorphisms. The results of this study reveal that significantly increased risks were found for the occurrence of DN in individuals with *GSTM1* genotype null. However, this same study does not observe correlation between the DN and the *GSTT1* genotype null or the presence of val alleles. Despite this, the genotype combination results indicate interaction between *GSTT1* null and *GSTM1* null, suggesting a possible summation in the deficiencies of these enzymes.

These findings differ from those found in a previous study by *Fujita* et al. (2000), where no associations between DN and genotype *GSTM1* null were observed. This study was performed with two groups of Japanese patients with or without diabetic nephropathy. Statistical analyses show that the frequency of the null genotype *GSTM1* was not significantly higher in the group of patients with nephropathy than in the group of patients without nephropathy, suggesting that the null *GSTM1* genotype does not contribute to the development of DN in this population [59].

More recent studies with the Romanian population suggest that the polymorphism of the *GSTP1* Ile105Val gene was associated with the risk of developing type 2 DM, but not with the risk of developing DN. For polymorphisms in the *GSTM1* and GSTT1 genes, the results did not indicate an increased risk of developing DM or DN [30].

Studies with the Brazilian population do not indicate an association of *GSTM1* deletion polymorphism with type 2 DM susceptibility. However, the *GSTM1* null and *GSTT1* null polymorphisms reveal an influence on some observed clinical parameters (blood glucose and blood pressure). This suggests that both polymorphisms may contribute to the clinical course of patients with type 2 DM [60].

On the other hand, other studies with the population of Central Brazil [61] suggest that individuals with null *GSTT1* polymorphism present an increased risk of approximately 2.9-fold for DN development. For the same population, no association of *GSTM1* null and DN was found. In this same study, the analysis of the influence of the deletion of *GSTT1* and *GSTM1* on clinical and biochemical changes did not indicate a significant association, and this suggests that the *GSTT1* null polymorphism may be associated with the risk of developing the disease, but not with the biochemical alterations analyzed.

6. Conclusion

Considering all the information described above, it is concluded that DM is among the main health concerns in the world. Hyperglycemia is the main characteristic of this pathology, and this unusual situation favors the imbalance between the reactive oxygen species and the antioxidant defense line produced by the individual. This condition is called oxidative stress and Glutathione and GSTs enzymes are fundamental for the reestablishment of redox homeostasis. The progression of diabetes and, consequently, prolonged exposure to this condition, favor the development of secondary complications of DM. DN is the main secondary complication that arises as result of DM.

Expression of polymorphic GST genes within several ethnic populations is remarkable. Some studies have suggested an association between genetic polymorphism of GSTs M1, T1 and P1 susceptibility to DM and its microvascular complications, and others do not. As the results are still scarce and inconsistent, more studies need to be done.

Acknowledgements

The authors would like to thank the National Council for Scientific and Technological Development (CNPq) (Grant: 448905/2014-0 to A.A.S.R.).

Author details

Laura Raniere Borges dos Anjos[1], Ana Cristina Silva Rebelo[5], Gustavo Rodrigues Pedrino[4], Rodrigo da Silva Santos[1,3] and Angela Adamski da Silva Reis[1,2]*

*Address all correspondence to: angeladamski@gmail.com

1 Laboratory of Molecular Pathology, Institute of Biological Sciences (ICB II), Federal University of Goiás (UFG), Goiânia, GO, Brazil

2 Department of Biochemistry and Molecular Biology, Institute of Biological Sciences (ICB II), Federal University of Goiás (UFG), Goiânia, GO, Brazil

3 Department of Natural Sciences (LEdoC), Special Academic Unit of Human Sciences, Federal University of Goiás (UFG), Goiás, GO, Brazil

4 Department of Physiology, Institute of Biological Sciences (ICB II), Federal University of Goiás (UFG), Goiânia, GO, Brazil

5 Department of Morphology, Institute of Biological Sciences (ICB III), Federal University of Goiás (UFG), Goiânia, GO, Brazil

References

[1] Milech et al. Guidelines Brazilian Society of Diabetes [Internet]. Available from: http://www.diabetes.org.br/profissionais/images/docs/DIRETRIZES-SBD-2015-2016.pdf [Accessed: 2017/12/10]

[2] International Diabetes Federation. Diabetes. In: Atlas of Diabetes. 8th edition. 2017 [Internet]. Available from http://www.diabetesatlas.org/ [Accessed: 2017/12/20]

[3] David JA, Rifkin WJ, Rabbani PS, Ceradini DJ. Review Article: The Nrf 2/Keap 1 /ARE Pathway and Oxidative Stress as a Therapeutic Target in Type II Diabetes Mellitus; 2017. DOI:10.1155/2017/4826724

[4] Gönül N, Kadioglu E, Aygün N, Özkaya M, Esat A, Karahalil B. The role of GSTM, GSTT1, *GSTP1*, and OGG1 polymorphisms in type 2 diabetes mellitus risk : A case – Control study in a Turkish population. Gene. 2012;**505**(1):121-127

[5] Raza H. Dual localization of glutathione S-Transferase in the cytosol and mitochondria: Implications in oxidative stress. Toxicity and Disease. NIH Public Access. 2011;**278**(22):4243-4251. DOI: 10.1111/j.1742-4658.2011.08358.x

[6] Pahwa S, Sharma R, Singh B. Role of glutathione S-Transferase in coronary artery disease patients with and without type 2 diabetes mellitus. 2017:5-8. DOI: 10.7860/JCDR/2017/23846.9281

[7] Halliwell B. Reactive species and antioxidants. Redox biology is a fundamental theme of aerobic life. *Plant Physiology*. 2006;**141**(2):312-322. DOI: 10.1104/pp.106.077073

[8] Pisoschi AM, Pop A. The role of antioxidants in the chemistry of oxidative stress: A review. European Journal of Medicinal Chemistry. 2015;**97**:55-74. DOI: 10.1016/j.ejmech.2015.04.040

[9] Forcados GE, Chinyere CN, Shu ML. Acalypha wilkesiana: Therapeutic and toxic potential. Journal of Medical and Surgical Pathology. 2016;**1**:122. DOI: 10.4172/jmsp.1000122

[10] Barbosa KBF, Costa NMB, Alfenas RCG, Paula SO, Minim VPR, Bressan J. Oxidative stress: Concept, implications and modulatory factors. Journal of Nutrition. 2010;**23**(4):629-643. DOI: 10.1590/S1415-52732010000400013

[11] Birben E, Sahiner UM, Sackesen C, Erzurum S, Kalayci O. Oxidative stress antioxidant Defense. Journal WAO. 2012. DOI: 10.1097/WOX. 0b013e3182439613

[12] Mc Cord JM. The evolution of free radicals and oxidative stress. The American Journal of Medicine. 2000;**108**:652-659. PMID: 10856414

[13] Matsui et al. Increased formation of oxidative DNA damage, 8-hydroxy-20-deoxyguanosine, in human breast cancer tissue and its relationship to *GSTP1* and COMT genotypes. Cancer Letters. 2000;**151**:87-95. PMID: 10766427

[14] Young IS, Woodside JV. Antioxidants in health and disease. Journal of Clinical Pathology. 2001;**54**:176-186. PMID: 11253127

[15] Harman D. Role of free radicals in aging and disease. Annals of the New York Academy of Sciences. 1992;**673**:126-141. PMID: 1485710

[16] Lee JD, Cai Q, Shu XO, Nechuta SJ. The role of biomarkers of oxidative stress in breast cancer risk and prognosis: A systematic review of the epidemiologic literature. Journal of Women's Health. 2017;**26**:467-482. DOI: 10.1089/jwh.2016.5973

[17] Tangvarasittichai S. Oxidative stress, insulin resistance, dyslipidemia and type 2 diabetes mellitus. World Journal of Diabetes. 2015;**6**:456-480. DOI: 10.4239/wjd.v6.i3.456

[18] Dalle-Donne I, Rossi R, Colombo R, Giustarini D, Milzani A. Biomarkers of oxidative damage in human disease. Clinical Chemistry. 2006;**52**:601-623. DOI: 10.1373/clinchem.2005.061408

[19] Rao AL, Bharani M, Pallavi V. Role of antioxidants and free radicals in health and disease. Advances in Pharmacology and Toxicology. 2006;**7**:29-38

[20] Lobo et al. Free Radicals, Antioxidants and Functional Foods: Impact on Human Health. Pharmacognosy Reviews 4.8. 2010:118-126. PMC. Web. 22 December 2017. DOI: 10.4103/0973-7847.70902

[21] Hayes J, Flanagan J, Jowsey I. Glutathione transferases. Annual Review of Pharmacology and Toxicology. 2005;**45**:51-88. DOI: 10.1146/annurev.pharmtox.45.120403.095857

[22] Huber PC, Almeida WP. Glutathione and related enzymes: Biological role and importance in processes pathological. Quimica Nova. 2008;**31**(5):1170-1179. DOI: 10.1590/S0100-40422008000500046

[23] Marí M, Morales A, Colell A, García-Ruiz C, Fernández-Checa JC. Mitochondrial glutathione, a key survival antioxidant. Antioxidants & Redox Signaling. 2009;**11**(11):2685-2700. DOI: 10.1089/ARS.2009.2695

[24] Wu G, Fang YZ, Yang S, Lupton JR, Turner ND. Glutathione metabolism and its implications for health. The Journal of Nutrition. 2004;**134**(3):489-492. PMID: 14988435

[25] Deponte M. Glutathione catalysis and the reaction mechanisms of glutathione-dependent enzymes. Biochimica et Biophysica Acta. 2013;**1830**(5):3217-3266. DOI: 10.1016/j.bbagen.2012.09.018

[26] Graczyk-Jarzynka et al. New insights into redox homeostasis as a therapeutic target in B-cell malignancies. Current Opinion in Hematology. 2017 Jul;**24**(4):393-401. DOI: 10.1097/MOH.0000000000000351

[27] Bjørklund G, Chirumbolo S. Role of oxidative stress and antioxidants in daily nutrition and human health. Nutrition. 2017;**33**:311-321. DOI: 10.1016/j.nut.2016.07.018

[28] Yang Y, Awasthi YC. Glutathione S-transferases as modulators of signal transduction. In: Taylor, Awasthi YC, editors. Toxicology of Glutathione Transferases. Boca Raton, Fl, USA: Francis CRC Press; 2006. pp. 205-230

[29] Orlewski J, Orlewska E. Effects of genetic polymorphisms of glutathione S- transferase genes (GSTM1, GSTT1, *GSTP1*) on the risk of diabetic nephropathy: A meta-analysis. Polskie Archiwum Medycyny Wewnętrznej. 2015;**125**:649-658. PMID: 26252359

[30] Stoian et al. Influence of GSTM1, GSTT1, and *GSTP1* Polymorphisms on Type 2 Diabetes Mellitus and Diabetic Sensorimotor Peripheral Neuropathy Risk. Disease Markers. 2015;**2015**:10 p. Article ID 638693. DOI:10.1155/2015/638693

[31] Barseem N, Elsamalehy M. Gene polymorphisms of glutathione S-Transferase T1/M1 in Egyptian children and adolescents with type 1 diabetes mellitus. Journal of Clinical Research in Pediatric Endocrinology. 2017 Jun;**9**(2):138-143. DOI: 10.4274/jcrpe.3690

[32] Lee et al. Gama-Glutamyltransferase and diabetes- a 4 years follow-up study. Diabetologia. 2003;**46**:359-364. DOI: 10.1007/s00125-003-1036-5

[33] Júnior Rover L, Hoehr NF, Vellasco AP. Antioxidant system involving the metabolic cycle of glutathione associated with electroanalytical methods in the evaluation of oxidative stress. Quim Nova. 2001;**24**(1, São Paulo). DOI: 10.1590/S0100-4042201000100019

[34] Barreiros ALBS, David JM. Oxidative stress: Relationship between generation of reactive species and defense of the organism. NOVA Chemicals. 2006;**29**(1):113-123

[35] Netto et al. Update on glycated hemoglobin (HbA1C) to assess glycemic control and to diagnose diabetes: Clinical and laboratory aspects. Jornal Brasileiro de Patologia e Medicina Laboratorial. February 2009;**45**(1):31-48. DOI:10.1590/S1676-24442009000100007

[36] Saha et al. Correlation between oxidative stress, nutrition, and cancer initiation. International Journal of Molecular Sciences. 2017. DOI: 10.3390/ijms18071544

[37] Baynes, Dominiczak HM. Glucose homeostasis and energetic metabolism: Diabetes mellitus. In: Medical Biochemistry, 4th Edition. Elsevier Saunders; 2014

[38] Brownlee M. Biochemistry and molecular cell biology of diabetic complications. Nature. 2001;**414**:813-820. DOI: 10.1038/414813a

[39] Schaan BD. The role of protein kinase C in the development of vascular complications of diabetes mellitus. Arquivos Brasileiros de Endocrinologia e Metabologia. December 2003;**47**(6). DOI: 10.1590/S0004-27302003000600006

[40] Idris I, Gray S, Donnelly. Protein kinase C activation: Isozyme-specific effects on metabolism and cardiovascular complications in diabetes. Diabetologia. 2001;**44**:659-673. DOI: 10.1007/s001250051675

[41] Nishikawa et al. Normalizing mitochondrial superoxide production blocks three pathways of hyperglycaemic damage. Nature. 2000;**404**:787-790. DOI: 10.1038/35008121

[42] Zimniak P. Substrate and reaction mechanism of GSTs. In: Taylor, Awasthi YC, editors. Toxicology of Glutathione Transferases. Boca Raton, Fl, USA: Francis CRC Press; 2006. pp. 71-102

[43] Giacco F, Brownlee M. Oxidative stress and diabetic complications. Circulation Research. 2010;**107**:1058-1070. DOI: 10.1161/CIRCRESAHA.110.223545

[44] Dandona et al. Oxidative damage to DNA in diabetes mellitus. Lancet. 1996;**347**:444-445. PMID: 8618487

[45] Williams et al. Glucoseinduced protein kinase C activation regulates vascular permeability factor mRNA expression and peptide production by human vascular smooth muscle cells in vitro. Diabetes. 1997;46:1497-1503. PMID: 9287052

[46] Rizvi S, Raza ST, Mahdi F. Association of genetic variants with diabetic nephropathy. World Journal of Diabetes. 2014;5(6):809-816. DOI: 10.4239/wjd.v5.i6.809

[47] Ceriello A, Testa R, Genovese S. Clinical implications of oxidative stress and potential role of natural antioxidants in diabetic vascular complications. Nutrition, Metabolism, and Cardiovascular Diseases. 2016;26:285-292. DOI: 10.1016/j.numecd.2016.01.006

[48] Sun L, Dutta RK, Xie P, Kanwar YS. Myo-inositol oxygenase overexpression accentuates generation of reactive oxygen species and exacerbates cellular injury following high glucose ambience: A new mechanism relevant to the pathogenesis of diabetic nephropathy. The Journal of Biological Chemistry. 2016;291:5688-5707. DOI: 10.1074/jbc.M115.669952

[49] Gluhovschi C, Gluhovschi G, Petrica L, Timar R, Velciov S, Ionita I, et al. Urinary biomarkers in the assessment of early diabetic nephropathy. Journal of Diabetes Research. 2016;2016:4626125. DOI: 10.1155/2016/4626125

[50] Mahmoodnia et al. An update on diabetic kidney disease, oxidative stress and antioxidant agents. Journal of Renal Injury Prevention. 2016. DOI: 10.15171/jrip.2017.30

[51] Filla LA, Edwards JL. Metabolomics in diabetic complications. Molecular BioSystems. 2016;12:1090-1105. DOI: 10.1039/C6MB00014B

[52] Oniki K, Umemoto Y, Nagata R, Hori M, Mihara S, Marubayashi T, Nakagawa K. Glutathione S-transferase A1 polymorphism as a risk factor for smoking-related type 2 diabetes among Japanese. Toxicology Letters. 2008;178:143-145. DOI: 10.1016/j. toxlet.2008.03.004

[53] Bid HK, Konwar R, Saxena M, Chaudhari P, Agarwal CG, Banerjee M. Association of glutathione S-transferase (GSTM1, T1 and P1) gene polymorphisms with type 2 diabetes mellitus in north Indian population. Journal of Postgraduate Medicine. 2010;56:176-181. DOI: 10.4103/0022-3859.68633

[54] Amer MA, Ghattas MH, Abo-Elmatty DM, Abou-El-Ela SH. Influence of glutathione S-transferase polymorphisms on type-2 diabetes mellitus risk. Genetics and Molecular Research. 2011;10:3722-3730. DOI: 10.4238/2011

[55] Di Pietro G, Magno LA, Rios-Santos F. Glutathione S-transferases: An overview in cancer research. Expert Opinion on Drug Metabolism & Toxicology. Feb 2010;6(2):153-170. DOI: 10.1517/17425250903427980

[56] Tiwari et al. Oxidative stress pathway genes and chronic renal insufficiency in Asian Indians with type 2 diabetes. Journal of Diabetes and its Complications. 2009;23:102-111. DOI: 10.1016/j.jdiacomp.2007.10.003

[57] Burden AC, McNally PG, Feehally J, Walls J. Increased incidence of endstage renal failure secondary to diabetes mellitus in Asian ethnic groups in the United Kingdom. Diabetic Medicine. 1992;9:641-645. PMID: 1511571

[58] Mostafa S. Evaluation of glutathione S-transferase P1 (*GSTP1*) Ile105Val polymorphism and susceptibility to type 2 diabetes mellitus, a meta-analysis. Excli Journal: Experimental and Clinical Sciences. 2017;**16**:1188-1197. DOI: 10.17179/excli2017-828

[59] Fujita et al. No association of glutathione S-transferase M1 gene polymorphism with diabetic nephropathy in Japanese type 2 diabetic patients. Renal Failure. 2000;**22**:479-486. DOI: 10901185

[60] Pinheiro et al. Evaluation of glutathione S-transferase GSTM1 and GSTT1 deletion polymorphisms on type – 2 diabetes mellitus risk. PLoS One;**8**(10):e76262. DOI: 10.1371

[61] Lima, RM. The role of the metabolic polymorphism of GSTM1 and GSTT1 in the susceptibility to diabetic nephropathy [Thesis]. 91 f. Dissertation (Master in Biology)—Federal University of Goiás, Goiânia. 2016

Glutathione and Central Nervous System

Chapter 5

The Role of Glutathione in Viral Diseases of the Central Nervous System

Juliana Echevarria Lima

Additional information is available at the end of the chapter

http://dx.doi.org/10.5772/intechopen.76579

Abstract

The function and physiology of the central nervous system (CNS) can be affected by of bacterial, fungal, protozoan, and viral infections. The neurological effects of viruses are associated with direct infections of structures of the CNS, the migration of infected leukocytes to the CNS, or/and the immune response to control the infection. In all these situations, we have reactive oxygen species (ROS) generation. ROS induces several cellular effects, including cell cycle progression, apoptosis, DNA damage, senescence, and neurodegeneration. The control of ROS involves the glutathione (GSH) balance, owing to antioxidant activity. Moreover, GSH is related with the transport of endogenous/exogenous molecules to extracellular medium by ABCC1/MRP1 activity. The depletion of GSH levels characterizes viral infections and associated-disease progression. Many studies correlated the GSH levels with immune response and suggest adding the glutathione replenishment to highly active antiviral treatment. Thus, it is important to review the relationship between the CNS, immune response, and GSH levels during neurological viral diseases.

Keywords: GSH, JC virus, CMV, HIV-1, HTLV-1, central nervous system, neurological viral diseases, ABCC1/MRP1, immune response

1. Introduction

There are many infectious pathogens that are etiologic agent of central nervous system (CNS) diseases, including the broad categories of bacteria, fungi, parasites, and virus. These infections are an important cause of morbidity and mortality in the world. The viral CNS infections are associated with meningitis and encephalitis development principally. However, the viral infections also are related with diseases in the CNS characterized by the presence of leukocyte infiltration and inflammation, inducing a progressive damage [1], such as the progressive multifocal leukoencephalopathy with the John Cunningham virus (JC), AIDS-related dementia complex observed in HIV-1-infected patients, neurodevelopmental sequelae (mental retardation, cerebral palsy, and sensorineural hearing loss) caused by congenital cytomegalovirus (CMV) infection or cerebral mass lesions in immunocompromised adults CMV-infected, and HTLV-1-associated myelopathy/tropical spastic paraparesis (HAM/TSP) that affects the human T-cell lymphotropic virus type 1 (HTLV-1)-infected individuals.

In normal oxidative metabolism, the free radical formation is expected. During the 1950s, researchers observed the occurrence of reactive oxygen species (ROS) during molecule irradiation with X-rays and as an effect of normal enzyme metabolic activity. They started to propose that the formation of oxygen free radicals induced tissues and cell damage [2]. At the same time, it was suggested that the mice treatment with glutathione (GSH) inhibited the animal deaths caused by X-ray irradiation [2, 3].

GSH is a tripeptide synthesized in all mammalian cells from the amino acid precursors L-glutamate, L-cysteine, and glycine, through the reactions catalyzed by γ-glutamylcysteine and GSH synthetase. Physiologically, 98% of intracellular glutathione is found in reduced form, and only 2% is detected under oxidized form (GSSH) or joined with other molecules [4]. Glutathione (GSH) has an important role in cellular physiology and metabolism, including antioxidant activity and induction of cellular proliferation [5]. Furthermore, the GSH-dependent antioxidant enzymes (glutathione peroxidase-1, glutathione reductase, glutathione S-transferase) cooperate and are interconnected reactions that eliminate ROS or controlled the redox state. Dysregulation of GSH synthesis was associated with many diseases, such as diabetes mellitus, cholestatic liver disease, endotoxemia, alcoholic liver disease, cancer, and neurodegenerative diseases. During aging the GSH content was decreased in the liver, lung, kidney, red blood cells, spleen lymphocytes, cerebral cortex, and cerebellum. This GSH concentration decline was related with the reduced expression of proteins involved in GSH synthesis. The GSH levels have been studied in Alzheimer's and Parkinson's diseases and others conditions [6]. However, the CNS is exposed to many situations that can induce a cell and tissue damage associated with ROS production. In this chapter, we will discuss some aspects of the balance of GSH levels and oxidative stress during viral infections in the CNS.

2. Intracellular levels of GSH in viral-infected cells and related cellular alterations

The JC virus is a double-stranded (ds) DNA virus from Polyomaviridae family. The mechanism of human-to-human transmission of the JC virus has not been established. It has been suggested that the ingestion of contaminated water and food represents the portal of entrance of this virus in human. The virus entry in the CNS goes through the blood-brain barrier (BBB), infecting the brain microvascular endothelium cells. The virus also infects B lymphocytes in the periphery that like a Trojan horse infiltrated the CNS in immunocompromised patients. There the virus infects oligodendrocytes and astrocytes [7]. JC virus is an etiologic agent of progressive multifocal leukoencephalopathy (PML), a demyelinating disease. Unfortunately, this disease is currently untreatable and fatal. The relationship between progressive multifocal leukoencephalopathy and GSH still remains unknown. Moreover, JC virus has also been related with CNS tumors, astrocytomas, glioblastomas, neuroblastomas, and medulloblastomas in immunosuppressed and non-immunosuppressed individuals [8]. However, GSH and GSH-related enzymes constitute an important mechanism of drug and multidrug resistance to glioblastomas, as described below [9].

CMV is a member of *beta*-herpesvirus subfamily, in the family Herpesviridae. It is the largest human herpesvirus, with a 230-kb ds DNA genome infection. Virus is spread from infected individual to noninfected individual by body fluids, such as urine, saliva, blood, tears, semen, and breast milk. In addition, a CMV-infected woman can pass the virus to her developing baby during pregnancy [10]. Congenital CMV infection causes serious neurodevelopmental sequelae, including mental retardation, cerebral palsy, and sensorineural hearing loss. CMV also is an increasingly important opportunistic pathogen in immunocompromised patients, inducing cerebral mass lesions. Antiviral therapy of children with symptomatic CNS congenital CMV infection is effective at reducing the risk of long-term disabilities [11].

In muscle cells the CMV infection induces ROS production minutes after entry. This phenomenon is associated to the virus life cycle. The increase in ROS levels activates the transcriptional factor NF-κB, leading transactivation of the viral genes and inducing the transcription of viral proteins [12]. On the other hand, the infection induces increased levels of GSH to control the ROS generation in vitro. This GSH augment is essential to produce the viral progeny. These data suggested that CMV infection coordinates conditions where ROS levels should be controlled and oxidative stress minimized [13]. However, the CMV infection in peripheral blood erythrocytes of pregnant women induces reduced of GSH and GSH peroxidase levels, leading an increase of H_2O_2 levels. These effects were associated with hemolytic anemia in pregnant women [14]. Although the CMV infection has been demonstrated in human brain cells in vitro, such as endothelial cells, astrocytes, neuronal cells, oligodendrocytes, and microglia [11], these studies did not investigate the role of GSH in CMV infection.

HTLV-1 and HIV were classified to the genus *Lentivirus* within the family of Retroviridae, subfamily Orthoretrovirinae. This virus infects leukocytes, which circulate in the blood and lymphatic vessels and may infiltrate in the spinal cord or brain, inducing a neurological diseases [7]. These viruses can be transmitted vertically from mother to child during transplacental transfer, delivery, or breastfeeding, by sexual contact and parenterally through the transfusion of the blood, organ transplant, and blood components or through contaminated needles.

HTLV-1 is the etiological agent of the adult T-cell leukemia/lymphoma and HTLV-1-associated myelopathy/tropical spastic paraparesis (HAM/TSP), a chronic progressive disabling disease characterized by demyelination, axonal loss, neuronal degeneration, and gliosis. The main site of neurodegeneration is the thoracic spinal cord; this leads to a slowly progressive spastic paraparesis with low back pain and bowel, urinary, and sexual dysfunction. The treatment consists in diminishing the symptoms, using corticosteroid therapy [15, 16]. It was demonstrated that Tax, a HTLV-1-viral protein, induces an increase in ROS generation, causing DNA damage and cellular senescence [17]. Moreover, it was observed that the persistence of the virus in infected cells involves mitochondrial ROS production modulated by viral protein p13 [18]. The CD4+ T lymphocytes are the main targets of HTLV-1 infection, but it has been shown that other leukocytes and glial cells are also infected [19–21]. The infected cells can migrate to the spinal cord and induced the HAM/TSP development. HTLV-1-infected individuals present a spontaneous T-lymphocyte proliferation. This phenomenon is related to the HTLV-1-proviral load and the persistence of the infection. The spontaneous proliferation induced by HTLV-1 infection depends on intracellular GSH levels. Using a GSH synthesis inhibitor, DL-Buthionine-[S,R]-sulfoximine (BSO), the spontaneous proliferation induced by HTLV-1 was impaired in peripheral blood mononuclear cells (PBMC) from infected donors. On the other hand, the GSH precursor induces an increase in mitogen-stimulated cellular proliferation in HTLV-1-infected individuals [19]. Thus, modulation of GSH levels could be proposed as a therapeutic target in HTLV-1-associated diseases.

HIV infection is associated with acquired immune deficiency syndrome (AIDS) development. Combinations of antiretroviral drugs administered as highly active antiretroviral therapy (HAART) reduced the AIDS mortality. However, since 1997 it has been described that 10–20% of virus-infected individuals present HIV-associated dementia [22]. Moreover, the HIV infection also causes mild neurocognitive disorder and demyelinating neuropathy with motor and sensory impairments. The treatment of neurological diseases in HIV-1-infected individuals is established based at the symptoms in association or not with HAART. After virus entry in the CNS, it infects and replicates in microglia that acquire inflammatory phenotype, inducing a neurological damages [7]. Several groups showed a GSH deficiency in the HIV-infected tissues. Initially, the studies demonstrated that low levels of GSH were related to the impairment of T-cell functions. In this work, the authors observed an increase of survival of AIDS patients after the oral treatment with N-acetylcysteine (NAC) administration, a precursor of GSH [22, 23]. In addition, the thiols level analysis in the cerebrospinal fluid (CSF) from HIV-infected patients indicated a significantly reduction in GSH and cysteinyl-glycine levels. Furthermore, the treatment of HIV-infected patients with S-adenosylmethionine, a precursor of homocysteine that is used in GSH synthesis, induced an increase of GSH levels in CSF [24]. Together, these findings suggested the importance of GSH modulation during HIV infection, but the pathway involved to alter the GSH levels remained unknown. The mechanism of neurodegeneration involves the viral protein, gp41. Neurons' death was observed when these cells were incubated in the presence of lentivirus lytic peptide (LLP-1) that expresses the carboxy terminal cytoplasmic domain of gp41 from HIV-1. In addition, the incubation the neuron cell lines with LLP-1 induced a decrease in GSH levels, mitochondrial membrane depolarization, and H_2O_2 production rapidly. The combination of GSH or NAC with LLP-1 prevented the mitochondrial membrane depolarization and cell death [25].

The development of neurological HIV disorders depends on virus entry in the CNS. To access the CNS, HIV virus particles and the infected cells induce the BBB disruption. HIV envelope protein gp120 and the regulatory virus protein, Tat, are involved in BBB breakdown. The incubation of endothelial brain cells with gp120 and Tat reduced the GSH intracellular levels and decreased GSH/GSSG ratio. These proteins also caused an increase in lipid peroxidation, suggesting that gp120 and Tat played an important role in BBB disruption by induction of oxidative cellular stress in endothelial cells [26].

The antioxidant response signals induce the activation of nuclear factor erythroid-derived 2-like-2 (Nrf2). It is a transcription factor that translocates into the nucleus and binds in the

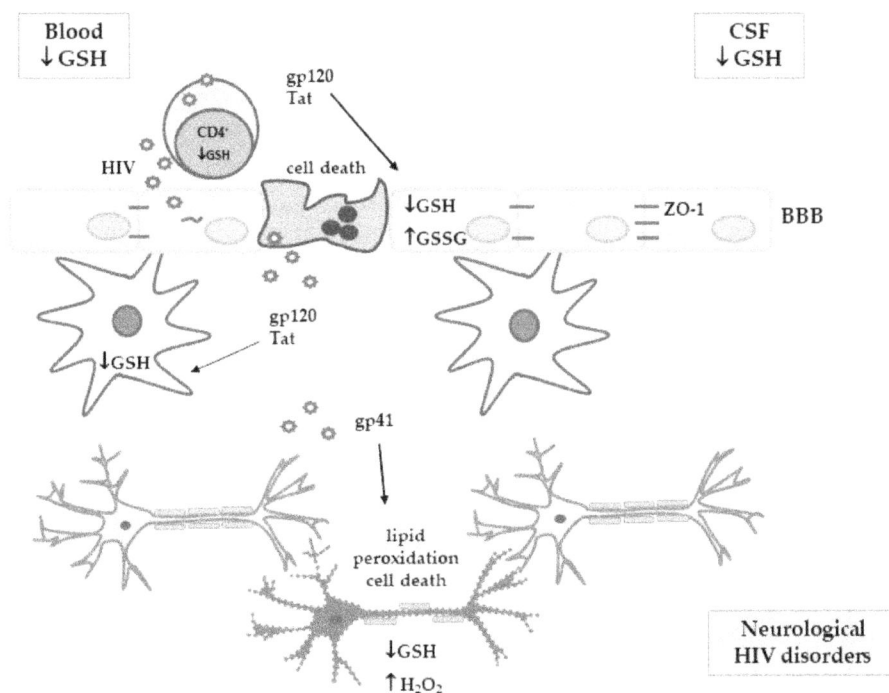

Figure 1. Levels of GSH during HIV-1 infection. In HIV-1-infected patients lower levels of GSH in the peripheral blood and CSF were observed. Alterations in intracellular levels of GSH cells were showed in vivo and in vitro. HIV proteins— gp120, Tat, and gp41—reduced intracellular levels of GSH in endothelial cells, astrocytes, and neurons. The effects of HIV-1-viral proteins also involved BBB breakdown, lipid peroxidation, and cell death.

promoter regions of detoxifying and antioxidants genes [27]. Viral protein Tat enhanced cellular expression of Nrf2 and its translocation into the nucleus. Nrf2 overexpression inhibited the Tat effects, reducing the intracellular ROS and increasing intracellular levels of GSH [28].

The effects of gp120 and Tat can be observed in vivo using a mice model. The administration of gp120 and Tat together or alone decreased GSH and GSH peroxidase brain levels. Animals also presented a reduction of tight junction protein ZO-1, suggesting other effects into BBB, and exhibited augment in lipid peroxidation in the brain [29]. In **Figure 1**, the direct effects of HIV infection in GSH levels and the consequences of this modulation are summarized.

3. HTLV-1 and HIV-1 infection and GSH active transport

GSH is related with the transport of endogenous and exogenous molecules to extracellular medium. GSH is a physiological substrate of ABCC1. Multidrug resistance-related protein 1 (ABCC1) transports several compounds in a GSH-dependent manner; its activity could be stimulated by the GSH intracellular levels. The members of the ABCC family are ATP-dependent efflux pumps, belonging to the ABC family of transport proteins, and they are also involved in resistance against anticancer drugs. ABCC1 is expressed in tumor cells [30] and normal tissues, such as the brain [31] and lymphocytes [32]. ABCC1 expression depends on Nrf2 activation and translocation to the nucleus [33].

It was already described in this charter; JC virus was detected in CNS tumors, such as glioblastomas. This brain tumor is highly proliferative and invasive and presents mechanisms of multidrug resistance (MDR). It was found that MDR glioblastoma cells displayed lower levels of endogenous ROS and high levels of GSH. On the other hand, the redox state disequilibrium or down modulation of GSH made these MDR cells more sensitive to chemotherapy [9]. In JC virus-infected glioblastoma cells, it is possible to find the same MDR feature. However, the influence of proteins from virus in MDR mechanisms expression remains unknown.

T lymphocytes CD4+ and CD8+ from HAM/TSP asymptomatic and symptomatic individuals presented a reduced ABCC1 expression and activity when compared to uninfected ones [34]. However, a lower ABCC1 expression was detected in CD4+ T lymphocytes from symptomatic patients. This result was directly correlated to the proviral load; a lower expression of ABCC1 was observed in patients with higher proviral load [34]. The pharmacological inhibition of ABCC induced a proliferation increase induced by mitogen of lymphocytes obtained from HTLV-1-infected individuals [19]. The expression and activity of ABCC1 transporter in BBB during HTLV-1 infection still remain unknown. It was suggested that dysregulations of ABC efflux transporters were implicated with the BBB breakdown during neurological diseases [35]. In infectious diseases this phenomenon can be involved in virus entrance in the CNS.

The incubation of astrocytes with gp120 enhanced the mRNA and protein levels of ABCC1. This effect was followed by the increase in substrate fluorescent or GSH transport and decreasing of GSSG efflux. Together these results suggested that the balancing of oxidative cellular status involves the increase in active GSH efflux to extracellular medium [36]. HIV protease inhibitors—ritonavir, indinavir, saquinavir, nelfinavir, and zidovudine—were described as ABCC1 substrate [35], suggesting that the overexpression of ABCC1 in infected cells makes these cells more resistant to chemotherapy.

4. Role of GSH in HIV and HTLV-1 immune response

T CD4+ lymphocyte differentiation involves the antigen-presenting cells (APCs) that display antigen complexed with major histocompatibility complex class II (MHC II) on their surfaces. The antigens are associated to MHC II molecule that interacts with T-cell receptor of T CD4+ lymphocytes, leading the antigen recognition and, subsequently, activation. T-CD4+ lymphocyte activation can generate some profiles (named Th), which depend on molecules present in the microenvironment. The cell phenotype is related with a group of cytokines and other immune products produced by T cell, generating inflammatory or anti-inflammatory cells. During viral infections the activation of inflammatory T-cell phenotype can be associated with virus eradication. However, in the CNS the exacerbation of inflammatory response is related with neurodegeneration [37]. Mice infected with the retroviral complex LP-BM5, a murine model of AIDS, presented GSH and/or cysteine reduction in lymphoid organs (spleen and lymph nodes). This GSH down modulation was followed by change in cytokine profile. The

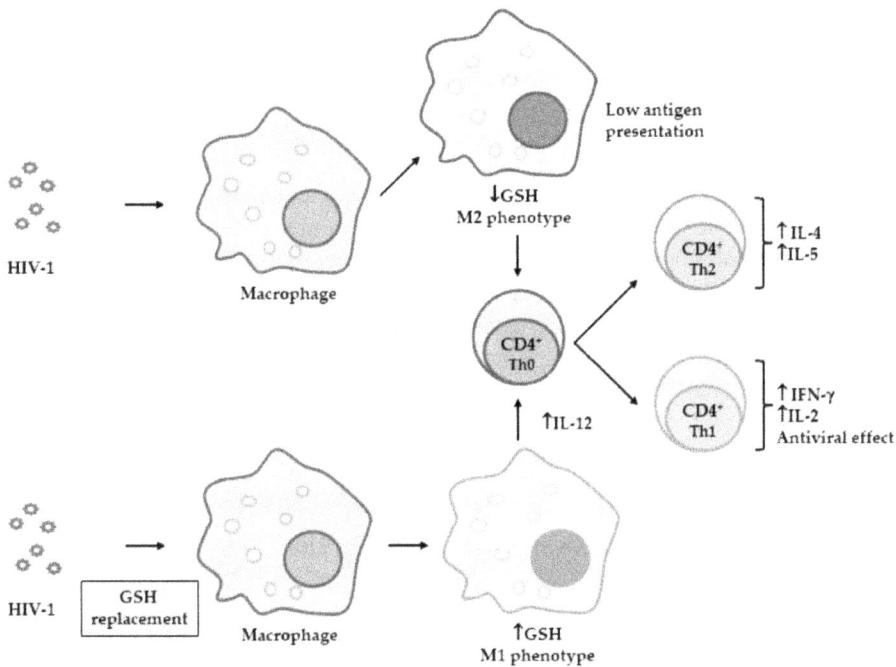

Figure 2. Role of GSH in HIV-1 immune response. The infection induced generation of M2 macrophages and T lymphocytes with Th2 phenotype. However, the GSH replacement led macrophages to M1 differentiation and CD4$^+$ lymphocyte secretion of Th1 cytokines.

infected mice exhibited a higher increase in interleukin (IL) IL-5, IL-4, and IL-2 than IL-12 and interferon-γ (IFN-γ), suggesting an important alteration in cytokine profile from Th1 to Th2 (**Figure 2**) [38].

Macrophages and dendritic cells are an important group of APCs. During infections macrophages can acquire specialized functional phenotypes. Macrophages classic activated are involved in inflammatory responses and are denominated M1. Macrophages alternative activated exhibit an antagonic inflammatory profile and named M2 [37]. Macrophages HIV-1 and LP-BM5 infected exhibited a decrease in GSH and cysteine intracellular levels. In addition, low intracellular levels of GSH were correlated with defective processing of antigens in APCs, indicating that GSH may be a critical factor in antigen processing [39]. During the LP-BM5 infection, macrophage polarization into alternative profile was observed, suggesting that M2 cells were driving the T-cell phenotype. LP-BM5-infected mice treatment with GSH replacement changed the macrophage polarization to M1 profile, inducing an increase in Th1 cytokine production and augmented antiviral response [38]. Thus, GSH modulation causes immune response phenotype alteration, leading to an important impact in virus elimination (**Figure 2**).

T lymphocyte CD8$^+$ is a cytotoxic T cell. They recognize the antigens through binding between TCR and MHC class I associated with antigen peptide. The control of viral infection is directly linked with efficiency of CD8$^+$ cytotoxic response [37]. The treatment with NAC induced an increase in surface activation molecule CD69 expression on unstimulated CD8$^+$ T lymphocytes obtained from HTLV-1-infected individuals. This result suggested that the increase in CD69 expression on CD8+ lymphocytes from HTLV-1 infected donors was correlated with an augmentation of GSH. Thus, increases in GSH levels could be beneficial to the activation of HTLV-1-specific CD8+ T cell and to the elimination of HTLV-1-infected cells [19].

The neurodegeneration is associated with decontrolled inflammatory responses into the CNS. Inflammatory cytokines induce nitric oxide (NO) and ROS production for innate immune cells and microglial cells. The incubation of microglia cells in the presence of viral protein gp120 was observed to increase in ROS production [36]. Besides, gp120 induces secretion of tumor necrosis factor-α (TNF-α) and monocyte chemoattractant protein-1 (MCP-1), leading to neuronal cell death, subsequently [40]. The inflammatory microenvironment reduces the glutamate uptake, inducing accumulation of this excitatory amino acid and excitotoxic neurodegeneration. Although, any study has not related the viral infection, GSH intracellular levels, and excitotoxic neurodegeneration, the literature suggested that antioxidant responses can prevent the neuron death directly or indirectly.

5. Effects of antiviral therapy in GSH levels in the CNS

The strategy used to treat children with symptomatic CNS congenital CMV infection and immunosuppressed individuals CMV-infected is based on doses of ganciclovir. This is an acyclic deoxyguanosine nucleoside analogue [41]. In vivo studies using mice model infected with CMV demonstrated that the treatment with ganciclovir reduced a viral load and TNFα levels. Moreover, the results suggested that antiviral therapy suppressed the oxidative damage by downregulation of malondialdehyde and upregulation of GSH levels in mice serum [42]. Unfortunately, the role of ganciclovir in CNS oxidative damage related with CMV infection remains unknown.

No antiviral treatment intervention exists for HTLV-1 infection. The HAM/TSP treatment is limited to symptomatic therapy. Usually, symptomatic patients are treated with corticosteroid pulse therapy. During last decades the antiviral therapy against HIV was improved, resulting in a significant reduction AIDS-related mortality and increasing HIV-infected patient survival. The highly active antiretroviral therapy (HAART) is started with the combination of two nucleoside analogue transcriptase reverse inhibitors and one non-analogue nucleoside transcriptase reverse inhibitor or protease inhibitor plus ritonavir-boosted. The analysis of T CD4+ lymphocytes obtained from the peripheral blood of HIV-1-infected patients showed an increase in GSH levels and decrease in GSSG levels during HAART at 1 year. In this study the patients received one protease inhibitor (indinavir or ritonavir) in combination with two nucleoside analogs (lamivudine plus zidovudine or plus stavudine), suggesting that the HAART ameliorates the oxidative alterations related with HIV-1 infection [43]. However, the effects of HAART on GSH levels may be different in other cell types. Human aortic endothelial cells pre-exposed to HAART produced higher levels of ROS than untreated cells after phorbol myristate acetate stimulation. After the HAART treatment, T-lymphocyte cell adhesion on human aortic endothelial cell monolayer increases significantly. However, the addition of NAC or GSH induced the inhibition of these effects, suggesting that the modulation of antioxidant levels activated the endothelium [44].

The first approved antiretroviral drug was zidovudine (AZT), a nucleoside reverse transcriptase inhibitor. The relationship between AZT and GSH has been studied since 1998. Mice treatment with AZT did not exhibit a significant decrease in GSH in total muscle homogenate, but the GSSG concentration increases, leading an increase in GSSG/GSH ratio. Furthermore, AZT treatment induces a skeletal muscle mitochondrial peroxide production [45]. Similar results were observed in monocytic cell lines incubated in the presence of AZT. The AZT treatment induced a significant reduction in GSH levels and destruction of mitochondria [46]. AZT is the antiretroviral drug with the best intracerebral penetration, however this substance virus resistance mutations in periphery and CNS [47]. The effects of AZT in GSH levels in the CNS have been remained unknown. Zang et al. demonstrated that mouse neuron exposure for short term to AZT did not present alteration in mitochondrial DNA levels. However, the results suggested that AZT long-term exposure caused deletion of mitochondrial DNA and neuron death [48]. Furthermore, AZT or the combination AZT plus indinavir (protease inhibitor) induces oxidative stress in human brain microvascular endothelial cells. These cells represent an important model to study BBB. The combination AZT plus indinavir induced an increase in ROS production, disruption in membrane mitochondrial potential, reduction in intracellular GSH levels, augment permeability of endothelial layer, leading cell death [49]. Together these results suggested that this antiretroviral therapy compromises the BBB and could be associated with HIV-1 neurological diseases.

These findings suggested that the replacement of GSH, reducing the oxidative stress in HIV-1-infected patients, is an interesting therapeutic approach. In some therapeutic strategies, to restore the GSH levels NAC or pro-GSH molecules in combination with HAART have been used. Moreover, the higher levels of GSH improve the antiviral immune response, collaborating in viral load reduction and in maintaining normal T-CD4+ lymphocyte count [50].

6. Discussion and conclusion

In this chapter we explore some aspects about neurodegenerative diseases associated with viral infection, GSH, and oxidative stress. Worldwide, many individuals are afflicted by JC, CMV, HTLV-1, and HIV-1 and develop some neurological diseases. However, studies that describe how the oxidative stress is involved in disease development remain insufficient. The oxidative stress in the CNS is associated to many neurodegenerative diseases. ROS, including reactive nitrogen species, are important mediator of brain and spinal cord damage. They are related with inflammation and mitochondrial and proteasomal dysfunction. The vulnerability of the CNS is associated with the higher consumption of oxygen than other tissues. Oxygen is important in ATP generation process, which is responsible for energy support used during normal CNS function. Physiological ROS levels are essential to neuronal functions, such as enhancing synaptic plasticity, long-term potentiation, and memory formation. However, the brain endogenous antioxidant defenses have not been enough to your demand. Moreover, the complexity of the cell composition of this tissue and the elevated oxygen levels corroborate to elevated capacity of the CNS in ROS production. All cellular macromolecules are susceptible to oxidative harm. ROS level elevation activates the detoxification and repair pathways.

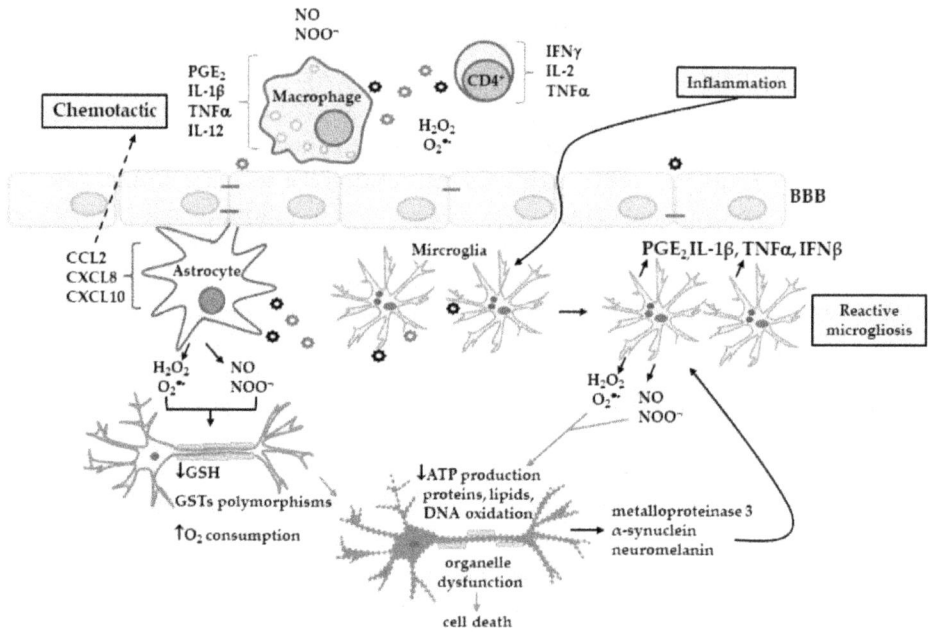

Figure 3. The imbalance of pro-oxidants induces oxidative stress and cell damage. The vulnerability of the CNS: ↓GSH and ↑O$_2$ consumption. Inflammation triggers microglia. Activated microglia releases inflammatory cytokines, ROS, and RNS. Microglia and astrocytes can be activated via pattern recognition receptors. During astrocyte activation, these cells released ROS, RNS, and chemokines. In this microenvironment neurons presented macromolecule oxidation, mitochondrial disruption, and, consequently, cell death.

The failure in these processes produces oxidation of proteins; lipids and DNA; consequently, organelle dysfunction; and after that neuronal damage. The critical organelle affected is the mitochondria, whose disruption induces reduction in ATP generation and apoptosis or necrosis [51]. As previously described the viral infection induced an increase in ROS production directly in CNS cells or indirectly by the infiltrated activated immune system cells, which use ROS release as mechanism to control the infection (**Figure 3**).

Glial cells (astrocytes and microglia) play important roles in maintaining CNS homeostasis through some processes, including reduction of oxidative stress. During neurodegenerative disorder glial cells release some factors to reestablish integrity and repair damaged cells. However, during the chronic inflammation, the glial activation causes an increase of ROS production and other neurotoxic mediators, leading a neuronal damage [52]. The principal cell type involved in CNS inflammation is the microglia. Microglia expresses some pattern recognition receptors that are engaged by pathogen-associated molecular patterns, triggering microglia activation. Activated microglia produced inflammatory mediators, such as prostaglandin E$_2$, interleukin-1β TNFα, ROS (peroxide—H$_2$O$_2$, superoxide—O$_2$•$^-$), and reactive species nitrogenous (RNS: NO; NOO- peroxynitrite). This phenomenon induces neuron damage. Damaged dopaminergic neurons release matrix metalloproteinase 3, α-synuclein, and neuromelanin that superactivated microglia, inducing reactive microgliosis, enhancing of the neurotoxicity-related mediators, such ROS (**Figure 3**). Moreover, ROS exerts an important effect on microglia as the second messenger, modifying inflammatory gene transcription and, consequently, amplifying the inflammation. During reactive microgliosis an increase of GSSG levels is observed. Studies in Parkinson have been suggested that dopaminergic neurons from substance nigra can be associated to GSH deficiency, becoming these cells more vulnerable to ROS [53].

The glutathione transferase (GST) activity can be relate to sensibility of neurons of ROS. GSTs conjugate molecules, including xenobiotics, with GSH, and then, this conjugated molecules can be actively transported to extracellular medium by ABCC transporters. Moreover, GSTs are involved in c-Jun N-terminal kinase (JNK) signaling pathway. ROS causes GSTs-JNK-c-Jun complex formation blocking JNK signaling pathway and preventing the events associated with this signaling cascade. GST gene polymorphisms have been identified and produce an important impact in enzyme activity. Some studies demonstrated that GST gene polymorphism carries have a positive correlation with brain cancer, Alzheimer's disease, and Parkinson's disease development risk [54]. The positive correlation between GST gene polymorphisms and hepatocellular carcinoma caused by hepatitis B virus chronic infection [55] and uterine cancer associated to human papilloma virus infection was described [56] (**Figure 3**). However, the relationship between GST gene polymorphisms and viral diseases of the CNS remains unknown.

The major studies relating viral infection and glia cells have been developed in HIV-1 model infection. Microglial cells exposed to HIV viral protein Nef release IFNβ. Then, IFNβ induces iNOS expression and NO production [57]. Furthermore, HIV-1 protein Tat induces NADPH oxidase activity in astrocytes. ROS produced by NADPH oxidase activity was related to chemokine (CCL2, CXCL8, and CXCL10) production, and it was inhibited by the treatment of astrocytes with NAC or NADPH oxidase inhibitors [58]. Together these results suggested that the HIV infection induces glia cell activation, ROS, and RNS which are directly involved in production of inflammatory mediators. The imbalance of prooxidants induces oxidative stress and cell damage (**Figure 3**).

The studies in viral diseases of the CNS have suggested an important link between GSH, immune response, and antiviral response. The findings indicated that the GSH replenishment can be used in highly active antiviral treatment. However, in asymptomatic HTLV-1 carries, this clinical approach should be the opposite result. The importance to study the relationship between GSH levels and viral neurological diseases is clear.

Acknowledgements

This work was supported by grants from Fundação Carlos Chagas Filho de Amparo à Pesquisa do Estado do Rio de Janeiro (FAPERJ) and Oncobiologia Program. The author thanks Renata Novaes and Raquel C. de Albuquerque to collaborate in GSH-related works.

Conflict of interest

The authors declare no conflict of interests.

Author details

Juliana Echevarria Lima

Address all correspondence to: juechevarria@micro.ufrj.br

Department of Immunology, Institute of Microbiology Paulo de Góes, Federal University of Rio de Janeiro, Rio de Janeiro, RJ, Brazil

References

[1] Riddell J, Shuman EM. Epidemiology of CNS infections. Neuroimaging Clinics of North America. 2012;**22**:543-556. DOI: 10.1016/j.nic.2012.05.003

[2] Gerschman R, Gilbert LD, Nye SW, Dwyer P, Fenn WO. Oxygen posing and X-irradiation: A mechanism in common. Science. 1954;**119**:623-626. DOI: 10.1126/science.119.3097.623

[3] Chapman WH, Sipe CR, Elitzholtz DC, Cronkite EP, Chambers FW Jr. Sulfhydryl-containing agents and the effects of ionizing radiations. 1. Beneficial effect of gluta-thione injection on X-ray induced mortality rate and weight loss in mice. Radiology. 1950;**55**:865-873. DOI: 10.1148/55.6.865

[4] Wang W, Ballatori N. Endogenous glutathione conjugates: Occurrence and biological functions. Pharmacological Reviews. 1998;**50**:335-356

[5] Lu SC. Glutathione synthesis. Biochimica et Biophysica Acta. 2013;**1830**:3143-3153. DOI: org/10.1016/j.bbagen.2012.09.008

[6] Liu H, Wang H, Shenvi S, Hagen TM, Liu RM. Glutathione metabolism during aging and in Alzheimer disease. Annals of the New York Academy of Sciences. 2004;**1019**:346-349. DOI: 10.1016/j.mam.2008.05.005

[7] Koyuncu OO, Hogue IB, Enquist LW. Virus infections in the nervous system. Cell Host & Microbe. 2013;**13**:379-393. DOI: 10.1016/j.chom.2013.03.010

[8] Noch E, Sariyer IK, Gordon J, Khalili K. JC virus T-antigen regulates glucose meta-bolic pathways in brain tumor cells. PLoS One. 2012;**7**:e35054. DOI: 10.1371/journal.pone.0035054

[9] Zhu Z, Du S, Du Y, Ren J, Ying G, Yan Z. Glutathione reductase mediates drug resis-tance in glioblastoma cells by regulating redox homeostasis. Journal of Neurochemistry. 2018;**144**:93-104. DOI: 10.1111/jnc.14250

[10] Murphy E, Rigoutsos I, Shibuya T, Shenk TE. Reevaluation of human cytomegalovirus coding potential. Proceedings of the National Academy of Sciences of the United States of America. 2003;**100**:13585-13590. DOI: 10.1073/pnas.1735466100

[11] Cheeran MC, Lokensgard JR, Schleiss MR. Neuropathogenesis of congenital cyto-megalovirus infection. Clinical Microbiology Reviews. 2009;**22**:99-126. DOI: 10.1128/CMR.00023-08

[12] Speir E, Shibutani T, Yu ZX, Ferrans V, Epstein SE. Role of reactive oxygen intermedi-ates in CMV gene expression and in the response of human smooth muscle cells to viral infection. Circulation Research. 1996;**9**:1143-1152. DOI: org/10.1161/01.RES.79.6.1143

[13] Tilton C, Clippinger AJ, Maguire T, Alwine JC. Human cytomegalovirus induces mul-tiple means to combat reactive oxygen species. Journal of Virology. 2011;**85**:12585-12593. DOI: 10.1128/JVI.05572-11

[14] Lutsenko MT, Andrievskaya IA, Kutepova OL. Morphofunctional characteristics of the glutathione cycle in peripheral blood erythrocytes of pregnant women with a history of cytomegalovirus infection exacerbation during gestation. Bulletin of Experimental Biology and Medicine. 2014;**157**:278-281. DOI: 10.1007/s10517-014-2544-7

[15] Leite AC, Silva MT, Alamy AH, Afonso CR, Lima MA, Andrada-Serpa MJ, Nascimento OJ, Araújo AQ. Peripheral neuropathy in HTLV-I infected individuals without tropical spastic paraparesis/HTLV-I-associated myelopathy. Journal of Neurology. 2004;**251**:877-881. DOI: 10.1007/s00415-004-0455-7

[16] Araujo AQ, Silva MT. The HTLV-1 neurological complex. Lancet Neurology. 2006;**5**:1068-1076. DOI: 10.1016/S1474-4422(06)70628-7

[17] Kinjo T, Ham-Terhune J, Peloponese JM Jr, Jeang KT. Induction of reactive oxygen spe-cies by HTLV-1 tax correlates with DNA damage and expression of cellular senescence marker. Journal of Virology. 2010;**84**:5431-5437. DOI: 10.1128/JVI.02460-09

[18] Silic-Benussi M, Cannizzaro E, Venerando A, Cavallari I, Petronilli V, La Rocca N, Marin O, Chieco-Bianchi L, Di Lisa F, D'Agostino DM, Bernardi P, Ciminale V. Modulation of mitochondrial K(+) permeability and reactive oxygen species production by the p13 protein of HTLV-1. Biochimica et Biophysica Acta. 1787;**2009**:947-954. DOI: 10.1016/j.bbabio.2009.02.001

[19] Novaes R, Freire-de-Lima CG, de Albuquerque RC, Affonso-Mitidieri OR, Espindola O, Lima MA, de Andrada Serpa MJ, Echevarria-Lima J. Modulation of glutathione intracel-lular levels alters the spontaneous proliferation of lymphocyte from HTLV-1 infected patients. Immunobiology. 2013;**218**:1166-1174. DOI: 10.1016/j.imbio.2013.04.002

[20] Watabe K, Saida T, Kim SU. Human and simian glial cells infected by human T-lymphotropic virus type I in culture. Journal of Neuropathology and Experimental Neurology. 1989;**48**:610-619. DOI: 10.1016/j.jneuroim.2016.08.012

[21] Gudo ES, Silva-Barbosa SD, Linhares-Lacerda L, Ribeiro-Alves M, Real SC, Bou-Habib DC, Savino W. HAM/TSP-derived HTLV-1-infected T cell lines promote morphological and functional changes in human astrocytes cell lines: Possible role in the enhanced T cells recruitment into CNS. Virology Journal. 2015;**12**:165. DOI: 10.1186/s12985-015-0398-x

[22] McArthur JC, McClernon DR, Cronin MF, Nance-Sproson TE, Saah AJ, St Clair M, Lanier ER. Relationship between human immunodeficiency virus-associated dementia and viral load in cerebrospinal fluid and brain. Annals of Neurology. 1997;**42**:689-698. DOI: 10.1002/ana.410420504

[23] Herzenberg LA, De Rosa SC, Dubs JG, Roederer M, Anderson MT, Ela SW, Deresinski SC, Herzenberg LA. Glutathione deficiency is associated with impaired survival in HIV disease. Proceedings of the National Academy of Sciences of the United States of America. 1997;**94**:1967-1972. DOI: 10.1073/pnas.94.5.1997

[24] Castagna A, Le Grazie C, Accordini A, Giulidori P, Cavalli G, Bottiglieri T, Lazzarin A. Cerebrospinal fluid S-adenosylmethionine (SAMe) and glutathione concentrations in HIV infection: Effect of parenteral treatment with SAMe. Neurology. 1995;**45**:1678-1683. DOI: org/10.1212/WNL.45.9.1678

[25] Sung JH, Shin SA, Park HK, Montelaro RC, Chong YA. Protective effect of glutathione in HIV-1 lytic peptide 1-induced cell death in human neuronal cells. Journal of Neurovirology. 2001;**7**:454-465. DOI: org/10.1080/135502801753170318

[26] Price TO, Ercal N, Nakaoke R, Banks WA. HIV-1 viral proteins gp120 and Tat induce oxidative stress in brain endothelial cells. Brain Research. 2005;**1045**:57-63. DOI: 10.1016/j.brainres.2005.03.031

[27] Wasserman WW, Fahl WE. Functional antioxidant responsive elements. Proceedings of the National Academy of Sciences of the United States of America. 1997;**94**:5361-5366. DOI: 10.1073/pnas.220418997

[28] Zhang HS, Li HY, Zhou Y, Wu MR, Zhou HS. Nrf2 is involved in inhibiting Tat-induced HIV-1 long terminal repeat transactivation. Free Radical Biology & Medicine. 2009;**47**:261-268. DOI: 10.1016/j.freeradbiomed.2009.04.028

[29] Banerjee A, Zhang X, Manda KR, Banks WA, Ercal N. HIV proteins (gp120 and Tat) and methamphetamine in oxidative stress-induced damage in the brain: Potential role of the thiol antioxidant N-acetylcysteine amide. Free Radical Biology & Medicine. 2010;**48**:1388-1398. DOI: 10.1016/j.freeradbiomed.2010.02.023

[30] Cole SP, Bhardwaj G, Gerlach JH, Mackie JE, Grant CE, Almquist KC, Stewart AJ, Kurz EU, Duncan AM, Deeley RG. Overexpression of a transporter gene in a multi-drug-resistant human lung cancer cell line. Science. 1992;**258**:1650-1654. DOI: 10.1126/science.1360704

[31] Regina A, Koman A, Piciotti M, El Hafny B, Center MS, Bergmann R, Couraud PO, Roux F. Mrp1 multidrug resistance-associated protein and P-glycoprotein expression in rat brain microvessel endothelial cells. Journal of Neurochemistry. 1998;**71**:705-715. DOI: 10.1046/j.1471-4159.1998.71020705.x

[32] Legrand O, Perrot JY, Tang R, Simonin G, Gurbuxani S, Zittoun R, Marie JP. Expression of the multidrug resistance-associated protein (MRP) mRNA and protein in normal peripheral blood and bone marrow haemopoietic cells. British Journal of Haematology. 1996;**94**:23-33. DOI: 10.1046/j.1365-2141.1996.d01-1776.x

[33] Hayashi A, Suzuki H, Itoh K, Yamamoto M, Sugiyama Y. Transcription factor Nrf2 is required for the constitutive and inducible expression of multidrug resistance-associated protein 1 in mouse embryo fibroblasts. Biochemical and Biophysical Research Communications. 2003;**310**:824-829. DOI: org/10.1016/j.bbrc.2003.09.086

[34] Echevarria-Lima J, Rumjanek VM, Kyle-Cezar F, Harab RC, Leite AC, dos Santos Ornellas D, Moralles MM, Araújo AQ, Andrada-Serpa MJ. HTLV-I alters the multidrug resistance associated protein 1 (ABCC1/MRP1) expression and activity in human T cells. Journal of Neuroimmunology. 2007;**185**:175-181. DOI: 10.1016/j.jneuroim.2007.01.008

[35] Qosa H, Miller DS, Pasinelli P, Trotti D. Regulation of ABC efflux transporters at blood-brain barrier in health and neurological disorders. Brain Research. 1628;**2015**:298-316. DOI: 10.1016/j.brainres.2015.07.005

[36] Ronaldson PT, Bendayan R. HIV-1 viral envelope glycoprotein gp120 produces oxidative stress and regulates the functional expression of multidrug resistance protein-1 (Mrp1) in glial cells. Journal of Neurochemistry. 2008;**106**:1298-1313. DOI: 10.1111/j.1471-4159.2008.05479.x

[37] Owen JA, Punt J, Stranford SA, Jones PP, Kuby J. Kuby Immunology. 7th ed. New York: W.H. Freeman; 2013

[38] Brundu S, Palma L, Picceri GG, Ligi D, Orlandi C, Galluzzi L, Chiarantini L, Casabianca A, Schiavano GF, Santi M, Mannello F, Green K, Smietana M, Magnani M, Fraternale A. Glutathione depletion is linked with Th2 polarization in mice with a retrovirus-induced immunodeficiency syndrome, murine AIDS: Role of proglutathione molecules as immunotherapeutics. Journal of Virology. 2016;**90**:7118-7130. DOI: 10.1128/JVI.00603-16

[39] Short S, Merkel BJ, Caffrey R, McCoy KL. Defective antigen processing correlates with a low level of intracellular glutathione. European Journal of Immunology. 1996;**26**:3015-3020. DOI: 10.1002/eji.1830261229

[40] Guo L, Xing Y, Pan R, Jiang M, Gong Z, Lin L, Wang J, Xiong G, Dong J. Curcumin protects microglia and primary rat cortical neurons against HIV-1 gp120-mediated inflammation and apoptosis. PLoS One. 2013;**8**:e70565. DOI: 10.1371/journal.pone.0070565

[41] James SH, Kimberlin DW, Whitley RJ. Antiviral therapy for herpesvirus central nervous system infections: Neonatal herpes simplex virus infection, herpes simplex encephalitis, and congenital cytomegalovirus infection. Antiviral Research. 2009;**83**:207-213. DOI: 10.1016/j.antiviral.2009.04.010

[42] Lv Y, Lei N, Wang D, An Z, Li G, Han F, Liu H, Liu L. Protective effect of curcumin against cytomegalovirus infection in Balb/c mice. Environmental Toxicology and Pharmacology. 2014;**37**:1140-1147. DOI: 10.1016/j.etap.2014.04.017

[43] Aukrust P, Müller F, Svardal AM, Ueland T, Berge RK, Frøland SS. Disturbed glutathione metabolism and decreased antioxidant levels in human immunodeficiency virus-infected patients during HAART—Potential immunomodulatory effects of antioxidants. The Journal of Infectious Diseases. 2003;**188**:232-238. DOI: 10.1086/376459

[44] Mondal D, Pradhan L, Ali M, Agrawal KC. HAART drugs induce oxidative stress in human endothelial cells and increase endothelial recruitment of mononuclear cells: Exacerbation by inflammatory cytokines and amelioration by antioxidants. Cardiovascular Toxicology. 2004;4:287-302. DOI: 10.1385/CT:4:3:287

[45] de la Asunción JG, del Olmo ML, Sastre J, Millán A, Pellín A, Pallardó FV, Viña J. AZT treatment induces molecular and ultrastructural oxidative damage to muscle mitochondria. Prevention by antioxidant vitamins. The Journal of Clinical Investigation. 1998;102:4-9. DOI: 1 0.1172/JCI1418

[46] Yamaguchi T, Katoh I, Kurata S. Azidothymidine causes functional and structural destruction of mitochondria, glutathione deficiency and HIV-1 promoter sensitization. European Journal of Biochemistry. 2002;269:2782-2788

[47] Blum MR, Liao SH, Good SS, de Miranda P. Pharmacokinetics and bioavailability of zidovudine in humans. The American Journal of Medicine. 1988;85:189-194

[48] Zhang Y, Song F, Gao Z, Ding W, Qiao L, Yang S, Chen X, Jin R, Chen D. Long-term exposure of mice to nucleoside analogues disrupts mitochondrial DNA maintenance in cortical neurons. PLoS One. 2014;9:e85637. DOI: 10.1371/journal.pone.0085637

[49] Manda KR, Banerjee A, Banks WA, Ercal N. Highly active antiretroviral therapy drug combination induces oxidative stress and mitochondrial dysfunction in immortalized human blood-brain barrier endothelial cells. Free Radical Biology & Medicine. 2011;50:801-810. DOI: 10.1016/j.freeradbiomed.2010.12.029

[50] Fraternale A, Paoletti MF, Casabianca A, Nencioni L, Garaci E, Palamara AT, Magnani M. GSH and analogs in antiviral therapy. Molecular Aspects of Medicine. 2009;30:99-110. DOI: 10.1016/j.mam.2008.09.001

[51] Patel M. Targeting oxidative stress in central nervous system disorders. Trends in Pharmacological Sciences. 2016;37:768-778. DOI: 10.1016/j.tips.2016.06.007

[52] Wang X, Michaelis EK. Selective neuronal vulnerability to oxidative stress in the brain. Frontiers in Aging Neuroscience. 2010;2:12. DOI: 10.3389/fnagi.2010.00012

[53] Block ML, Zecca L, Hong JS. Microglia-mediated neurotoxicity: Uncovering the molecular mechanisms. Nature Reviews Neuroscience. 2007;8:57-69. DOI: 10.1038/nrn2038

[54] Allocati N, Masulli M, Di Ilio C, Federici L. Glutathione transferases: Substrates, inihibitors and pro-drugs in cancer and neurodegenerative diseases. Oncogene. 2018;7:8-23. DOI: 10.1038/s41389-017-0025-3

[55] Yu MW, Yang SY, Pan IJ, Lin CL, Liu CJ, Liaw YF, Lin SM, Chen PJ, Lee SD, Chen CJ. Polymorphisms in XRCC1 and glutathione S-transferase genes and hepatitis B-related hepatocellular carcinoma. Journal of the National Cancer Institute. 2003;95:1485-1488. DOI: 10.1093/jnci/djg051

[56] Tew KD, Townsend DM. Glutathione-s-transferases as determinants of cell survival and death. Antioxidants & Redox Signaling. 2012;17:1728-1737. DOI: 10.1089/ars.2012.4640

[57] Mangino G, Famiglietti M, Capone C, Veroni C, Percario ZA, Leone S, Fiorucci G, Lülf S, Romeo G, Agresti C, Persichini T, Geyer M, Affabris E. HIV-1 myristoylated Nef treatment of murine microglial cells activates inducible nitric oxide synthase, NO_2 production and neurotoxic activity. PLoS One. 2015;**10**(6):e0130189. DOI: 10.1371/journal.pone.0130189

[58] Youn GS, Cho H, Kim D, Choi SY, Park J. Crosstalk between HDAC6 and Nox2-based NADPH oxidase mediates HIV-1 Tat-induced pro-inflammatory responses in astrocytes. Redox Biology. 2017;**12**:978-986. DOI: 10.1016/j.redox.2017.05.001